Two Manuals Combined into One Book

Marine Combat Stress

And

Military Training for Urban Operations

Department of Defense

Pentagon
Publishing

PREFACE

Combat Stress is the mental, emotional or physical tension, strain, or distress resulting from exposure to combat and combat-related conditions. Controlling combat stress is a command responsibility. In terms of Service members lost from action and reduced performance, combat stress seriously affects mission accomplishment. It is a leader's responsibility to take action to strengthen Service members' tolerance to combat stress and manage it in his or her unit.

Combat stress reactions are the result of exposure to the same conditions during military actions that cause physical injury and disease in battle or its immediate aftermath, and many combat stress reactions occur in persons who are also wounded or ill with disease. Rates of combat stress casualties vary greatly, with higher ratios during lengthy periods of intense combat. In Okinawa 1945, during a peak month of battle, the combat stress casualties among Marine Forces were reported as high as one for every two wounded in action (WIA). Under less lengthy periods, as suggested by data acquired from the Israeli Defense Forces fighting in Lebanon 1982, the ratio of combat stress casualties to WIA in small units can be as high as one to one. In the past, we have generally suffered as many as one battle stress casualty for every three to five WIA in heavy fighting. However, highly trained units with strong leadership and high esprit de corps have fewer combat stress casualties.

While this manual focuses on combat-induced stress reactions, it is important to emphasize that "combat stress" is not restricted only to combat, but may also arise from combat-like conditions present during military operations other than war. In an area of operations characterized by continuous action and high danger, our forces may experience high rates of stress casualties unless small-unit leaders are trained and prepared to manage stress.

This publication is written to inform small-unit leaders of stress characteristics and management techniques in order to *prevent, reduce, identify,* and *treat* combat stress reactions in the Service member's own unit to the maximum extent possible. A significant part of training is learning to control and cope with stress. Leaders must learn to cope with their own stress and then assist junior personnel in managing their stress. The application of combat stress management techniques helps conserve fighting strength and provides one more step toward achieving success.

Table of Contents

Chapter 2. Combat Stress Prevention, Management, and Control

Chapter 3. Command Leadership Actions

Chapter 4. Sleep Deprivation

Appendices

Chapter 1

Combat Stress Identification

1001. INTRODUCTION

The Marine Corps' success as a fighting force is dependent on leadership that maintains a balanced focus between mission accomplishment and troop welfare. The small-unit leader is the key to building and maintaining high unit morale and peak efficiency. He achieves this in part by knowing his troops and understanding their strengths and weaknesses. To maintain that same level of morale and efficiency in combat, the small-unit leader must understand how to recognize, prevent, and even personally contend with reactions to combat stress when it occurs in his unit. If a condition accounted for as many casualties in combat and the condition was at least partially preventable, the prudent combat leader would be interested in knowing more about it. Combat stress reaction(s), also called battle fatigue, is that condition. It has the potential to disable the most courageous Service member and influence the success or failure of a unit in accomplishing its mission.

1002. HISTORY

During the 1942-45 period in the European Theater, there was a ratio of one combat stress casualty for every three WIA. In a month of especially horrible, continuous fighting in Okinawa in 1945, the 6th Marine division had one stress casualty for every 1.8 WIA. However, the airborne divisions in Europe never had more than one for ten WIA, and usually less, even though they experienced very high casualties in some battles.

To accomplish a mission successfully, planners must use some guidelines to estimate losses from combat stress. What are the reasons for the tremendous range between high and low battle stress casualty rates? While the answer to this question is complex, it is clear that better-trained troops have fewer killed and wounded, and proportionally fewer stress casualties. Service members—especially leaders—can learn to recognize the symptoms and prevent or reduce the disruptive effects of combat stress.

1003. REACTIONS TO COMBAT STRESS

Service members exposed to danger experience physical and emotional reactions that are not present under more tranquil circumstances. Some reactions sharpen abilities to survive and win; other reactions may produce disruptive behaviors and threaten individual and unit safety. These adverse behaviors are collectively called *combat stress reaction*. The operative word is "behaviors." People in combat experience a range of emotions, but their behavior influences immediate safety and mission success.

Combat and combat-related military missions can also impose combinations of heavy physical work; sleep loss; dehydration; poor nutrition; severe noise, vibration, and blast; exposure to heat, cold or wetness; poor hygiene facilities; and perhaps exposure to infectious diseases, toxic fumes or substances. These, in combination with other influences—such as concerns about problems back home—affect the ability to cope with the perception of danger, and diminish the skills needed to accomplish the mission. Environmental stressors often play an important part in causing the adverse or disruptive combat stress reaction behaviors. The leader must work to keep each Service member's perception of danger balanced by the sense that the unit has the means to prevail over it. The leader must keep himself and his unit working at the level of stress that sustains performance and confidence. When troops begin to lose confidence in themselves and their leader, adverse stress reactions are most likely to occur.

It is important for the small-unit leader to recognize these adverse behaviors at the onset in order to intervene promptly for the safety and benefit of individual Service members and the unit. These behaviors may take many forms and can range from subtle to dramatic. Any Service member who shows persistent, progressive behavior that deviates from his baseline behavior may be demonstrating the early warning signs and symptoms of a combat stress reaction. Trying to memorize every possible sign and symptom is less useful to prompt diagnosis than to keep one simple rule in mind: *Know your troops, and be alert for any sudden, persistent or progressive change in their behavior that threatens the functioning and safety of your unit.*

1004. OBSERVING AND RECOGNIZING COMMON REACTIONS TO COMBAT STRESS

Ranges of fatigue, fear, anxiety, and depression affect most Service members in combat and in some military operations other than war. Mild stress reaction may be signaled by changes in behavior and only be discernible by the person himself or by close comrades. The unit leader and medical personnel depend on information from the Service member or his comrades for early recognition of combat stress reactions to provide prompt and appropriate help. Table 1-1 lists some mild stress reactions.

Table 1-1. Mild Stress Reactions.

PHYSICAL	EMOTIONAL
Trembling	Anxiety, indecisiveness
Jumpiness	Irritability, complaining
Cold sweats, dry mouth	Forgetfulness, inability to concentrate
Insomnia	Nightmares
Pounding heart	Easily startled by noise, movement, and light
Dizziness	Tears, crying
Nausea, vomiting or diarrhea	Anger, loss of confidence in self and unit
Fatigue	
"Thousand-yard" stare	
Difficulty thinking, speaking, and communicating	

Severe stress reactions may prevent the individual from performing his duties or create a concern for personal safety or the safety of others. A variety of more serious reactions or warning signs are listed in Table 1-2. These do not necessarily mean that the person must be relieved from duty, but warrant immediate evaluation and help by the small-unit leader.

Table 1-2. Severe Stress Reactions.

PHYSICAL	EMOTIONAL
Constantly moves around	Talks rapidly and/or inappropriately
Flinches or ducks at sudden sound and movement	Argumentative; acts recklessly
Shakes, trembles	Indifferent to danger
Cannot use part of body (hand, arm leg) for no apparent physical reason	Memory loss
Inability to see, hear or feel	Stutters severely, mumbles or cannot speak at all
Is physically exhausted; cries	Insomnia; severe nightmares
Freezes under fire or it totally immobile	Sees or hears things that do not exist
States vacantly, staggers or sways when standing	Has rapid emotional shifts
Panics, runs under fire	Socially withdrawn
	Apathetic
	Hysterical outbursts
	Frantic or strange behavior

Fatigue

The most common stress reactions include:

- Slow reaction time.
- Difficulty sorting out priorities.
- Difficulty starting routine tasks.
- Excessive concern with seemingly minor issues.
- Indecision, difficulty-focusing attention.
- A tendency to do familiar tasks and preoccupation with familiar details. These reactions may reach a point where the person becomes very passive, or wanders aimlessly.
- Loss of initiative with fatigue and exhaustion.

Muscular Tension

Headaches, Backache

Muscular tension increases strain on the scalp and spine and often leads to headaches, pain, and cramps.

Inability to Relax

Prolonged muscular tension wastes energy and leads to fatigue and exhaustion. Muscles must relax periodically to enable free blood flow, waste product flushing, and nutrient replenishment.

Shaking and Tremors

Mild Shaking

During incoming rounds, the individual may experience mild shaking. This symptom appears and disappears rapidly and is considered a normal reaction to conditions of great danger.

Marked or Violent Shaking

A common post-battle reaction, marked or violent shaking can be incapacitating if it occurs during the action. If shaking persists long after the precipitating stimulus ceases, or if there was no stimulus, the individual should be checked by medical personnel.

Perspiration

It is normal to experience either mild or heavy sweating or sensations of chilliness under combat stress.

Digestive and Urinary Systems

Nausea and Vomiting

"Butterflies in the stomach" is a common stress feeling. Vomiting may occur as a result of an extreme experience like that of a firefight, shelling, etc., or in anticipation of danger.

Loss of Appetite

Appetite loss may result as a reaction to stress. It becomes a significant problem if rapid weight loss occurs, or the person does not eat a sufficiently balanced diet to keep his muscles and brain supplied for sustained operations.

Abdominal Distress

Acute abdominal pain ("knotted stomach," "heartburn") may occur during combat. Persistent and severe abdominal pain is a disruptive reaction, and may indicate a medical condition.

Frequent Urination

Frequent urination may occur, especially at night.

Incontinence

During extremely dangerous moments, inability to control bowel and/or bladder functions may occur. Incontinence is embarrassing, but it is not abnormal under these circumstances.

Circulatory and Respiratory Systems

Heart Palpitations

Rapid heartbeat, a sense of pressure in the chest, occasional skipped beats, and sometimes chest pains are common with anxiety or fear. Very irregular heartbeats need to be checked by medical personnel.

Hyperventilation

Hyperventilation is identified by rapid respiration, shortness of breath, dizziness, and a sense of choking. It is often accompanied with tingling and cramping of fingers and toes. Simple solutions are increased exercise and breathing with a paper bag over the nose and mouth; or breathing slowly using abdominal muscles (called "abdominal breathing").

Faintness and Giddiness

These reactions occur in tandem with generalized muscular weakness, lack of energy, physical fatigue, and extreme stress. Brief rest should be arranged, if possible.

Sleep Disturbance

Difficulty Falling Asleep

Sometimes a Service member who has experienced intense battle conditions often cannot fall asleep even when the situation permits, or, when he does fall asleep, he frequently wakes up and has difficulty getting back to sleep.

Nightmares

Terror dreams, battle dreams, and nightmares of other kinds cause difficulty in staying asleep. Sleep disturbances in the form of dreams are part of the coping process. This process of *working through* combat experiences is a means of increasing the level of tolerance of combat stress. The individual may have battle-related nightmares or dreams that a close relative (spouse, parent) or another person important in his life has been killed in the battle. As time passes, the nightmares usually occur with less intensity and less frequency. In some cases, a Service member, even when awake, may re-experience the memory of the stressful experience as if it were recurring (called a "flashback"). This is usually triggered by a smell, sound or sight, and is not harmful as long as the Service member realizes it is only a memory and does not react inappropriately or feel overwhelmed. However, if it happens frequently or is very distressing, help should be sought from the chaplain or medical.

Restless Sleep

When a person is asleep, the sleep is not restful sleep if the person is constantly being half-wakened by noise, movement, or other stimuli. Heavy snoring often indicates poor quality sleep. The individual wakes up as tired as when he went to sleep. Finding a more comfortable position, away from distractions, can help.

Excessive Sleep

Individuals exhibiting a need for excessive sleep may be exhibiting symptoms of combat stress; however, excessive sleep is also a sign of substance abuse or depression (Persistent insomnia is a more common indicator of possible depression.)

Visual/Hearing Problems and Partial Paralysis

Stress-related blindness, deafness, loss of other sensations, and partial paralysis are not true physical injuries, but physical symptoms that unconsciously enable the individual to escape or avoid a seemingly intolerably stressful situation. These symptoms can quickly improve with reassurance and encouragement from comrades, unit medical personnel or battalion physician. If they persist, the physician must examine the individual to be sure there is not a physical cause; for example, laser range finders can cause temporary or partial blindness, and nearby explosions can cause ear damage. Individuals with these physical conditions are unaware of the causative relationship with their inability to cope with stress. They are honestly concerned with their physical symptoms and want to get better. They are willing to discuss them, and do not mind being examined. This is contrary to "malingerers" faking a physical illness, who are often reluctant to talk, or over-dramatize their disability and refuse an examination.

Visual problems include blurred vision, double vision, difficulty in focusing, or total blindness. Hearing problems include inability to hear orders and/or nearby conversations or complete deafness occurs. Paralysis or loss of sensation is usually confined to one arm or leg. Prickling sensations or rigidity of the larger joints occur. However, temporary complete immobility (with normal breathing and reflexes) can occur. If these reactions do not recover quickly with immediate reassurance, care must be taken in moving the casualty to medical for an evaluation to avoid making a possible nerve or spinal cord injury worse.

Bodily Arousal

Not all emotional reactions to stress are necessarily negative. For example, the body may become aroused to a higher degree of awareness and sensitivity.

Threat

In response to threat, the brain sends out chemicals arousing the various body systems. The body is ready to fight or take flight. The alerting systems of the experienced combat veteran become finely tuned, so that he may ignore loud stimuli that pose no danger (such as the firing of nearby friendly artillery). However, he may awaken from sleep at the sound of an enemy mortar being fired and take cover before the round hits. The senses of vision and smell can also become very sensitive to warning stimuli. The Service member may instantly focus and be ready to react.

Hyperalert

Hyperalert refers to being distracted by any external stimuli that might signal danger, and overreacting to things that are, in fact, safe. The hyperalert Service member is not truly in tune with his environment, but is "on a hair trigger." He is likely to over-react, misinterpret reassuring information as threats, and react without adequate critical thinking. Consequences can range from firing at an innocent noise to designating as hostile an innocent target.

Startle Reactions

Startle reactions are part of an increased sensitivity to minor external stimuli (on-guard reactions). Leaping, jumping, cringing, jerking, or other forms of involuntary self-protective motor responses to sudden noises are noted. The noises are not necessarily very loud. Sudden noise, movement, and light cause startle reactions; for example, unexpected movement of an animal (or person) precipitates weapon firing.

Anxiety (Fear of Death, Pain, and Injury)

Fear of death, pain, and injury causes anxiety reactions. After witnessing the loss of a comrade in combat, a Service member may lose self-confidence and feel overly vulnerable or incapable. The death of a buddy leads to serious loss of emotional support. Feelings of "survivor guilt" are common. The survivors each brood silently, second-guessing what they think they might have done differently to prevent the loss. While the Service member feels glad he survived, he also feels guilty about having such feelings. Understanding support and open grieving shared within the unit can help alleviate this.

Irritability

Reaction

Mild irritable reactions range from angry looks to a few sharp words, but can progress to more serious acts of violence. Mild irritability is exhibited by sharp, verbal overreaction to normal, everyday comments or incidents; flare-ups involving profanity; and crying in response to relatively slight frustrations.

Explosion of Aggressive Behavior

Sporadic and unpredictable explosions of aggressive behavior (violence) can occur with little or no provocation. For example, a

Service member tries to pick a fight with another Service member. The provocation may be a noise, such as the closing of a window, an accidental bumping or normal, verbal interaction.

Short Attention Span

Persons under stress have short attention spans. They find it difficult to concentrate. Short attention span causes a Service member to have difficulty following orders. The Service member does not easily understand what others are saying. The person has difficulty following directions, aiding others or performing unfamiliar tasks.

Depression

People respond to stress with protective defensive reactions against painful perceptions. Emotional dulling or numbing of normal responsiveness is a result. The reactions are easily observed changes from the individual's usual self.

Low Energy Level

The observer notes decreased effectiveness on the job, decreased ability to think clearly, excessive sleeping or difficulty falling asleep, and chronic tiredness. Emotions such as pride, shame, hope, grief, and gratitude no longer matter to the person.

Social Withdrawal

A Service member is less talkative than usual, shows limited response to jokes or cries. He is unable to enjoy relaxation and companionship, even when the tactical situation permits.

Change in Outward Appearance

If the Service member is in a depressed mood, he may be observed to exhibit very little body movement and an almost expressionless (mask-like) face.

Substance Abuse

Some Service members may attempt to use substances such as alcohol or other drugs as a means of escaping combat stress. The use of substances in a combat area makes some Service members less capable of functioning on the job. They are less able to adapt to the tremendous demands placed on them in combat.

Loss of Adaptability

Less common reactions include uncontrolled emotional outbursts such as crying, yelling, or laughing. Some Service members may become withdrawn, silent, and try to isolate themselves. Uncontrolled reactions can appear singly or in combination with a number of other symptoms. In this state, the individual may become restless, unable to keep still, and move aimlessly about. He may feel rage or fear, which he demonstrates by aggressive acts, angry outbursts or irritability.

Disruptive Reactions

Service members with disruptive, combat stress reactions—

- **Cannot function on the job.** In some cases, stress produces symptoms often associated with head injuries. For example, the person may appear dazed and may wander around aimlessly. He may appear confused and disoriented, and exhibit either a complete or partial memory loss. Service members exhibiting this behavior should be removed from duties until the cause for this behavior can be determined.

- **Compromise their own safety.** In a desperate attempt to escape the danger that has overwhelmed him, an individual may panic and become confused. The term *panic run* refers to a person rushing about without self-control. In combat, such a Service member easily compromises his safety and could possibly get killed. His mental ability becomes impaired to the degree that he cannot think clearly or follow simple commands. He stands up in a firefight because his judgment is clouded and he cannot understand the likely consequences of his behavior. He loses his ability to move and seems paralyzed.
- **Compromise the safety of others.** If panic is not quelled early, it can easily spread to others. A person in panic is virtually out of control and needs to be protected from himself. More than one person may be needed to exert control over the individual experiencing panic. However, it is also important to avoid threatening actions, such as striking him.

Referral to the Chaplain, Medical Officer or Combat Stress Control Team

Although the more serious or warning behaviors described in the preceding paragraphs usually diminish with help from comrades and small unit leaders, and time, some do not. An individual usually improves when able to get warm food, rest and an opportunity to share his feelings with comrades or small unit leader. If the symptoms endanger the individual, others or the mission, or if they do not improve within a day or two, or seem to worsen, get the individual to talk with the unit chaplain or medical officer. Access to mental health/combat stress control specialists may be sought, if available. **Do not** wait too long to see if the experience is better with time. Specialized training is not required to recognize severe stress reactions. The small-unit leader can usually determine if the individual is not performing his duties normally, not taking care of himself, behaving in an unusual fashion, or acting out of character.

1005. SUICIDE AWARENESS

Some behaviors and symptoms previously described are not only signs of stress reaction, but can also signal potential suicide risks. Service members must be ever vigilant for the signs and signals of a potential threat of suicide given by their fellow Service members. Appendix A provides useful tools for identifying these signals.

Chapter 2

Combat Stress Prevention, Management, and Control

2001. PREVENTING AND MANAGING COMBAT STRESS

The same leadership skills that apply to troop welfare and warfighting can effectively reduce or prevent combat stress reactions. Leaders should take preventive actions and address stress symptoms as they appear. Ignoring the early warning signs can increase the severity of stress reactions. Positive action to reduce combat stress also helps Service members cope with normal, everyday situations and makes them less likely to experience harmful combat stress reactions. Table 2-1 lists stress management techniques.

2002. STRESS-REDUCTION TECHNIQUES FOR LEADERS

To reduce stress, the leader should—

- Lead by inspiration, not fear or intimidation.
- Initiate and support stress management programs.
- Provide information to focus stress positively.
- Ensure each Service member has mastered at least two stress coping (relaxation) techniques, a slow one for deep relaxation and a quick one for on the job.

Table 2-1. Stress Management Techniques.

Assure every effort is made to provide for the troops' welfare.
Instill confidence in each Service member and his equipment, unit, and leadership.
Be decisive and assertive; demonstrate competence and fair leadership.
Provide sleep and/or rest, especially during continuous operations, whenever possible.
Ensure sleep for decisionmaking personnel.
Set realistic goals for progressive development of the individual and team.
Systematically test the achievement of these goals.
Recognize that battle duration and intensity increase stress.
Be aware of environmental stressors such as light level, temperature, and precipitation.
Recognize that individuals and units react differently to the same stress.
Learn the signs of stress in yourself and others.
Recognize that fear is a normal part of combat stress.
Rest minor stress casualties briefly, keeping them with their unit.
Be aware of background stress sources prior to combat; e.g., family concerns and/or separation, economic problems.
Provide an upward, downward, and lateral information flow to minimize stress due to a lack of communication.

Table 2-1. Stress Management Techniques (Continued).

Practice stress control through cross-training, task allocation, tasks matching, and task sharing.
Look for stress signs and a decreased ability to tolerate stress.
Practice and master stress-coping techniques.
Face combat stress; it is unhealthy to deny the stresses of combat.

- Look out for Service members' welfare.
- Communicate with Service members personally to learn of stressors and detect signs of stress.
- Understand that stress in response to threatening or uncertain situations is normal.
- Create a spirit to win under stress.
- Act as role model for self-control of stress reactions.

Realistic training is the primary stress-reduction technique. It assures Service members' maximum confidence in their skills and belief that their leaders are doing their best for them. Since the basic necessities of life assume even greater importance on the battlefield, leaders should:

- Ensure personnel are properly trained.
- Ensure training includes understanding of combat stress and how to deal with it.
- Place welfare of subordinates before personal welfare, but keep them capable.
- Ensure Service members get as much rest as possible.

- Ensure the best possible shelters are available.
- Keep Service members well supplied with food, water, and other essentials.
- Provide mail, news, and information avenues.
- Provide the best medical, logistical, and other support.
- Maintain high morale, unit identity, and esprit de corps.
- Keep the same unit members together.
- Assure that experienced unit members take care of and teach new members.

The unit's encouragement and support of Service members' efforts to cope with stress have a decisive effect. Unit actions can determine if Service members will endure combat exposure and accomplish their mission. Stress-coping indoctrination is part of every unit's combat training and is followed by a program of action. Controlling stress requires practice. Programs are tailored to individual units and improve with experience. Coping with stress is practiced under conditions as similar to combat as possible. A unit's ability to cope with stress in combat depends primarily on how rigorously and realistically training has been conducted.

2003. PERFORMANCE DEGRADATION PREVENTIVE MEASURES

Every Service member, team, and unit must learn to effectively sustain performance in continuous operations. This requirement applies especially to leaders. While it is an important ingredient, the determination to endure does not ensure effectiveness. Gaining the required capability goes beyond a high level of proficiency in combat skills and technical specialties. It means learning to identify the adverse conditions of continuous operations, cope with them, and overcome their effects. It also means learning how to slow the rate of performance degradation. Units, leaders, and personnel must prepare and execute plans and train to sustain performance. Adverse conditions progressively degrade Service member effectiveness. Fortunately, long-term remedies exist for slowing the rate of performance decline. These remedies, which must be introduced prior to combat, include safety, food intake, combat load, and physical fitness.

Safety

Safety, which encompasses such factors as using proper lifting techniques and staying alert and careful, is influenced by fatigue. Overtired Service members are more vulnerable to injury than those who are rested. After 72 hours of continuous combat, the tendency to seek shortcuts is very strong, and accident rates increase 50 percent. Fatigue affects all military systems, but it is especially hazardous when weapon systems are involved. Catastrophic accidents can occur when fatigued (and under-experienced) crews man weapon systems. Ways to safeguard Service members include developing and following safety standing operating procedures and increasing supervision during extended operations.

Food Intake

If Service members are too busy, stressed or tired to eat adequate rations during continuous operations, their caloric intake will be reduced. This may lead to both physical and mental fatigue and degraded performance. For example, in accidents judged to involve aviator fatigue, there is some indication that before the accidents occurred, the pilots had irregular eating schedules or missed one or more meals. In field tests done by the U.S. Army Natick Research, Development, and Evaluation Center, Natick,

MA, some personnel subsisting solely on meals, ready to eat (MREs) lost weight over just a few weeks. Leaders must encourage troops to eat all of the main items in their MREs, not just the candy, to get balanced nutrition. In various other field tests conducted during continuous operations, it was discovered that meals were frequently delivered late or missed altogether. Although the relationship between performance and nutrition is not clear, eating regularly is important in continuous operations, as well as providing hot meals at assigned times or when the workload has been sustained boosts morale.

Good nutrition is important. An inadequate diet degrades performance, reduces resistance to disease, and prolongs recuperation from illness and injury. When food and water become available in continuous operations, leaders must provide and maintain a supply of food that has the nutritional value commensurate with the physical activity and stress of battle. They must remind and encourage Service members to eat and drink properly. The excitement, stress, and rapid pace of events associated with field preparations can cause Service members to forget to drink liquids. Thus, they enter the early part of the field scenario inadequately hydrated. Dehydration may result, especially if the early scenario calls for assault of a position or rapid air/land deployment. Contributing to developing dehydration is the relative lack of moisture in MREs and other packets. In addition, Service members experiencing dehydration lose their appetite and reduce their food intake. This, in combination with dehydration, leads to degraded performance. Leaders must reemphasize drinking regimens to ensure that Service members are properly hydrated going into battle. Leaders must remind Service members to drink liquids in both hot and cold climates and must monitor fluid intake. If personnel drink only when thirsty, they will become dehydrated.

Combat Load

In combat, the load carried by a Service member significantly exceeds optimum recommended weights. In the case of a light infantry Service member, the combat load is as much as double the recommended load. Physical conditioning cannot compensate for this degree of excess. Service members tire faster and, in continuous combat, recovery from fatigue becomes more time-consuming.

When the Service member must carry excessive amounts of equipment, the effects of stress and lack of rest are magnified. Employing a load echelonment concept must be considered to ease the strain on Service members. In this concept, the unit separates an individual's equipment into two loads—fighting and existence. As the unit closes the objective, the heavier existence load is dropped and the Service member continues with the lighter fighting load.

Physical Fitness

Good physical conditioning delays fatigue, builds confidence, and shortens recovery times from illness and injury. It also prepares individuals to better cope with the physiological demands of stress. Service members in top physical condition can better control their internal physiological functions, which will improve their overall performance. Physical fitness—including aerobic fitness, muscular strength, and endurance—must be developed in all Service members to strengthen their ability to rebound from exhaustion. Aerobic fitness increases work capacity and the ability to withstand stress. While feelings of depression and moodiness accompany tiredness, aerobically fit Service members are affected less than those unfit. The ability to quickly recover from physically strenuous workloads is maintained by smart physical training, performed consistently and routinely. However, there is no evidence that good physical conditioning significantly reduces normal sleep requirements nor compensates for the deleterious impact of sleep deprivation on cognitive functioning. Sleep deprivation is discussed in greater detail in chapter 4.

2004. EFFECTIVE LEADERSHIP

The effective leader in combat is competent and reliable. He knows his job without question, and he can be counted on to do it regardless of the situation or circumstances.

Effective small-unit leadership reduces the impact of stress in several ways. Leaders understand the sources of combat stress and reactions to them. In addition, leaders manage stress problems to keep them from spreading throughout the organization by implementing the following actions:

- Continue mission performance; focus on immediate mission.
- Expect Service members to perform assigned duties.
- Remain calm, directive, and in control at all times.
- Let Service members know their reactions are normal and there is nothing seriously wrong with them.
- Keep Service members productive (when not resting) through recreational activities, equipment maintenance, and training to preserve perishable skills.
- Ensure Service members maintain good personal hygiene.
- Ensure Service members eat, drink, and sleep as soon as possible.
- Let the Service members express their thoughts. Do not ignore or make light of expressions of grief or worry. Give practical advice and put emotions into perspective.

A battalion commander in World War II made the following address to his Marines just before a new offensive:

> I know as well as you do that the going has been tough. Maybe I sweat it more than you do because I have more to worry about. But you've done a fine job, and I'm proud of every single man in my outfit. I assure you that everything possible will be done to give you the best available support, and I will not order you to attack unless I'm confident that you have a real chance to succeed . . . . The harder we fight now, the sooner we can finish this mess and get back to living the way we want.

That speech demonstrated a common sense approach in dealing with stress. Had this commander preached about the advantages of democracy over fascism or the evils of Hitler, his words would have fallen on deaf ears. Instead, he seized on the strong motivators of his troops: their *respect* for him as a leader, their desire to continue to be worthy of his respect, and their desire to share in the unit's accomplishments.

Generally, confidence is based on the technical and tactical competence exhibited by leaders and Service members. Since confidence is one of the strongest defenses against stress, its development is a major goal for every military leader. To gain confidence, Service members must believe in themselves, their equipment, other unit members, and their training. Above all, they must believe in their leaders' competence. Each of these beliefs is instilled and reinforced at every opportunity.

The following comments were made by a Marine colonel reflecting on his experience as a company commander in Vietnam.

> A feeling of helplessness will overtake men when they realize that they have been put in an untenable position. When the situation is so bad that the men cannot fight back, they are susceptible to fear. The

situation can be critical, but as long as the men can fight back, fear normally will not overtake them. The solution to this problem is: don't put your men in untenable positions. Granted, in every battle some men in a unit will find themselves in a situation where they can't fight back. This is inevitable especially in an attack. Proper use of fire team, squad, platoon, and company formations, as well as intelligent use of terrain, minimizes the danger of a large portion of the unit being pinned down to the extent where the men cannot return fire. If you are tactically and technically proficient, use common sense, apply the principles of war, and employ the firepower available to your unit, untenable situations can be avoided. Simply stated, know your job!

Those comments provide an insight and sensitivity to problems related to combat stress through small-unit leadership. His instincts told him how important it was to exhibit solid leadership qualities in order to allow his troops to operate at maximum efficiency. A unit builds confidence, esprit, integrity and cohesion when the *leaders know their jobs*.

2005. COMBAT STRESS MANAGEMENT BUILDING BLOCKS

Confidence in Leaders

Leaders must demonstrate effective leadership to earn their subordinates' loyalty and trust. Leaders are responsible for—

- Committing the unit to missions commensurate with abilities.
- Planning operations carefully and thoroughly.
- Preparing the unit to accomplish the mission.
- Leading and guiding the unit to mission accomplishment.

- Showing consistent good leadership that convinces subordinates their leaders know best what should be done, how it should be done, who should do it, and how long the task should take. Authority accompanies leadership beyond the automatic authority given by military rank and position. Authority and respect are earned based on confidence in a leader's ability to guide the unit to success.

Confidence in Training

Training helps Service members develop the skills required to do their jobs. Confidence is the result of knowing they have received the best possible training for combat, and are fully prepared. This confidence results from the following:

- Realistic training that ends with successful mastery.
- Relevance of training to survival and success on the integrated battlefield.
- Refresher and cross training.
- Systematic individual and collective training.

Confidence in Unit

Each Service member in a unit needs to become confident of the other unit members' competence. Individuals must stay and train together to gain that personal trust. Unless absolutely necessary, teams should not be disbanded or scrambled. Subunits in the same larger unit should have the same Standing Operating Procedures and training standards, so members can fit in quickly if teams have to be cross-leveled or reorganized after casualties occur. Confidence in the unit leads to feelings of security, which in turn allows members to sleep and positively focus stress. In combat, the unit must receive each member's highest commitment to unit loyalty. Mission accomplishment is the unit's highest priority.

Confidence in Equipment

Service members who learn to operate and maintain assigned equipment develop confidence in their ability to employ it. This, in combination with an individual's belief in his personal capabilities, raises overall confidence in fighting ability.

Cohesion and Morale

Good unit cohesion and morale offset the negative effects of combat stress. The foundation for any stress reduction program includes trust and confidence in the following:

- Fellow Service members.
- Competence and fairness of the unit leaders.
- Unit's technical abilities and military power.
- Equipment.
- Personal combat ability.
- Sense of support from the civilian community.
- Personal spiritual well-being.

Physical Conditioning

A strong relationship exists between physical stamina and the ability to resist combat stress. Good physical conditioning has physical and psychological benefits. Rigorous physical conditioning helps protect against the stress of continuous operations. A regular program of physical fitness to increase aerobic endurance, muscular strength, and flexibility is essential to combat readiness. As physical conditioning improves, Service members feel better about themselves, have greater confidence in each other, and their stress is reduced.

Unit training includes regular physical conditioning. This increases the members' tolerance to all types of stressors. The program is geared to the unit's combat mission, and exercises are tailored to the environment where the unit operates. The pace, length, and types of runs, road marches, and other activities are commensurate with the unit's need. Light infantry units need more demanding, longer road marches than maintenance units. Activities also include team athletics, which capitalize on cohesion-building aspects as well as physical benefits. The benefits of such a program include developing endurance through aerobic exercises, enhancing strength through weight training, and deprivation/physical stress training.

Family Care

Service members entering combat with financial worries or family problems risk breaking down under the additional stress of combat. Even positive but unfinished changes on the home front, such as a recent marriage or parenthood, can distract the Service members' focus on combat missions with worries that they will not live to fulfill their new responsibilities at home.

Leaders must be aware of this risk and assist members in handling personal matters before deployment. Pre-deployment and post-deployment family briefings should be conducted, and programs established to assist families before, during, and after deployment. When Service members know their families are cared for, they are better able to focus on their combat duties.

Coping with Individual Stress

Stress pushes the body to its limits and causes tension; relaxation reverses this process. Coping with personal stress is essential. Stress-coping skills are incorporated into unit training early, and command emphasis is placed on practicing them. Service members receive a block of instruction on stress-coping techniques, then are given supervised time each duty day to practice them quietly. After 3 or 4 weeks of practice, most Service members relax easily and quickly, even under highly stressful conditions. They will be able to naturally control stomach fluttering, heart rate, blood pressure, and stress.

The stress-coping exercises include deep breathing, muscle relaxation, and cognitive exercises. Deep breathing is the simplest to learn and practice; the others require longer instruction and more practice time. On request, the Combat Stress Control (CSC) team will provide instructional materials and assistance.

Deep-Breathing Exercise

This consists of slow, deep inhaling, which expands the chest and abdomen, holding it for 2 to 5 seconds, then exhaling slowly and completely, which pushes out the used air. This can be done for five breaths as a quick, mind-clearing exercise, or continuously to promote sleep. Abdominal or diaphragmatic breathing (making the stomach move the air, rather than the upper chest) is especially effective for stress control and, with practice, can be done simultaneously with task that require full attention.

Muscle Relaxation Exercises

These are more complex. They generally consist of concentrating on various muscle groups and the tensing and relaxing of limbs to relax the entire body. Quick versions for use in action consist of tensing all muscles simultaneously, holding for 15 seconds or more, then letting them relax and "shaking out the tension." Deep relaxation versions start in the feet and work up (or start in the head and work down), body part by body part (muscle group by muscle group), tensing and then relaxing each in turn, while noticing how each part feels warm after it relaxes.

Cognitive Exercises

These consist of self suggestion (positive self-talk); imagery (imagine being fully immersed in a deeply relaxing setting); rehearsal (imagine performing the stressful or critical task under pressure and doing it perfectly); and meditation (clearing the mind of all other thoughts by focusing on every breath and silently repeating a single word or phrase). These techniques involve creating positive mental images that reduce the effects of stressful surroundings, redirecting mental focus and learning to detach from stress. Service members are encouraged to practice stress management techniques and discuss their use in combat.

CAUTION

It is important not to use deep relaxation techniques at times when you need to be alert to dangers in your surroundings. Practice the quick relaxation techniques so you can use them automatically without distraction from the mission.

2006. CRITICAL EVENT DEBRIEF AND UNIT LEADER'S AFTER ACTION REVIEW

NOTE: On the subjects of Critical Event Debrief/After-Action Review, U.S. Army personnel will follow the guidance provided in FM 22-51, *Leaders' Manual for Combat Stress Control*, chapter 6.

Critical Event Debrief

A Critical Event Debrief (CED) is a structured group process designed to mitigate the impact of a critical event and to accelerate normal recovery of those personnel involved. The CED is normally conducted by a team composed of trained members, (medical officers, chaplains, mental health professionals, trained unit members). A CED's main value is to quickly restore unit cohesion and readiness to return to action, through clarifying what actually happened and clearing up harmful misperceptions and misunderstandings. It may also reduce the possibility of long term distress through sharing and acceptance of thoughts, feelings, and reactions related to the critical event. Ideally, CEDs are conducted 24 to 72 hours after the occurrence of the event, away from the scene and separate from any operational debriefing. The typical CED lasts from 2 to 3 hours. Situations that warrant a CED include:

- Death of unit member
- Death or suffering of noncombatants (especially women and/or children)
- Handling of the dead, management of carnage or the sight of devastation during disaster work
- Serious friendly fire incident

- Situation involving a serious error, injustice or atrocity
- Terrorist attack.

After Action Review

Although a CED would be warranted following such traumatic events as those listed, if access to CED trained professional team is not available, small-unit leaders can modify their After Action Review (AAR) to assist in identifying the level of stress reaction of unit members. The AAR or "hotwash" is a routine practice used by leaders to debrief post-mission operations. Accordingly, it provides a familiar, non-threatening forum for leaders to identify the levels of stress reaction experienced by the unit members using the information in paragraph 1004. The leader's AAR may be sufficient to restore unit readiness for further action. It may also identify personnel who need immediate or later referral for chaplain or mental health/CSC help for combat stress-related symptoms. In those instances when access to CED trained facilitators is not available, the unit leader should incorporate the following questions in the AAR:

Fact Phase
Ask unit members to describe the event from their individual perspectives. What was each member's specific role in the event?

Thought Phase
What were each member's first thoughts at the scene (or when the incident was first heard of)?

Reaction Phase
What was the worst thing about the event? What was thought and/or felt?

Symptom Phase
Describe probable thinking, physical, and emotional responses both at the scene and a few days afterward.

Training Phase
Relay information regarding stress reactions and what can be done about them. If prepared handouts are available, distribute them. Include points of contact within the unit.

Wrap-up Phrase
Reaffirm positive things. Summarize. Be available and accessible. Debriefing team/unit leaders together decide which individuals need more help or referral.

2007. EFFECTIVE COMBAT STRESS CONTROL PROGRAM

An effective stress management program starts with early planning, continues during employment, and extends beyond the return home. The CSC program is comprised of three phases: pre-deployment, deployment and combat, and post combat.

Phase 1: Predeployment

During pre-deployment, some stressors facing Service members include long working hours, preparation for training, fear of the future, family worries, and anxiety about the unit's readiness. Signs of poor coping include insomnia, increased use of alcohol, marital problems, increased bickering in the unit, irritability, and suicidal feelings. The most important preparatory steps to take in the predeployment phase are to:

- Conduct unit training and mission rehearsals.
- Prepare for changed sleep schedules and jetlag.

- Attend to task assignments and allocations.
- Conduct equipment and supply maintenance checks.
- Attend to personal and family matters. (Call the Family Service Center.)
- Integrate new members into the unit positively and actively.
- Welcome "significant others" (not just entitled beneficiaries) in the Family Support network information tree.
- Brief as much information about the operation as possible, consistent with operations security measures.
- Actively familiarize the unit members with the stressors and horrors they may encounter (for example, watch videotapes from network television showing suffering civilians or atrocities), but always followed immediately with what the unit will do in those situations.
- Arrange for pre-deployment training and education, especially for refresher training of stress reduction techniques from local mental health professionals, a Special Psychiatric Rapid Intervention Team (SPRINT) or CSC team, if available.

Unit Training

Because unit leaders have combatted the stressors associated with garrison living and peacetime training, they have learned to know their Service members and what affects their performance. All Service members should believe unit training has prepared them thoroughly for combat. Stress is reduced if the training improves belief in themselves, and the unit's ability to fight and win. They are assured training is designed and developed to meet combat requirements, provides the ability to fight successfully, and that supporting units receive the same realistic training.

Realistic mission rehearsal helps desensitize Service members against potential combat stressors. For example, wearing and realistically training in protective gear is important. By doing so in pre-deployment training, Service members become less distressed in combat. The unit should be exposed to extensive drills wearing protective gear during live-fire exercises before employment. Such training has the two-fold, stress-reduction benefit of building confidence and preparing Service members for combat.

Leaders who provide their Service members with advance knowledge about the enemy prepare them for the stress created during deployment and initial enemy encounter. It is important during such training to talk realistically about enemy strengths and weaknesses as well as those of their own units. While inspirational pep talks are also important at this time, they should not include biased, inaccurate information. Leaders earn trust and respect if their troops perceive them as accurate, dependable sources of information.

Physical Fitness Training

In the time-urgent work of preparing for deployment, leaders assure that physical fitness is sustained, either by the work involved or by selective unit physical training exercises. New unit members who are insufficiently fit receive special training.

Stress-Coping Skills Training

During preparation for deployment, the leaders direct the unit to practice stress coping and relaxation techniques, and are positive role models by demonstrating use of these techniques. If necessary, the chaplain and mental health professional personnel available to the units can provide additional training.

Sleep Discipline

Before deployment, unit leaders must consider fatigue and sleep loss occurring during combat. The enforcement of work and rest schedules begins early in pre-deployment training. During continuous operations, fatigue caused by lack of sleep is a major source of stress. Breaks in combat are irregular, infrequent, and unscheduled. Extended sleep is unlikely. *Sleep logistics* is emphasized such that sleep and rest are allocated or supplied like rations, water, equipment, and ammunition. Sleep discipline training addresses the following points:

- A unit-specific work-rest-sleep plan is developed and practiced.
- The unit leader or commander is included in the allocation of sleep and rest time, as lack of sleep will impair his judgment and decision-making skills as much as those of his subordinates.
- The plan allows Service members at least 5 hours of uninterrupted sleep, ideally between 2400 and 0600, every 24 hours. Persons receiving only 5 hours per 24 hours over a period of several days will accumulate a significant sleep debt.
- Sleep priority is assigned to those whose judgment and decision-making are critical to mission accomplishment.
- If received frequently, 1 hour of sleep or even 15-minute naps help, but "slow mental starting" upon wake up can result.
- Relaxation exercises complement sleep schedules. These exercises are used as an alternative to regular sleep or as an aid to help Service members rest under difficult circumstances.

If preparation is made for jetlag, its effects are minimal. The ideal method is to seclude units up to several weeks before deployment, gradually changing work, eat, and rest cycles to coincide with the destination's schedule. Although rapid deployment does not permit this method, there are ways to help make the transition to a new environment, such as changing to new sleep cycles collectively versus individually; sleeping up to 12 hours before operation start; ensuring leaders receive rest; and adjusting diet to meet energy needs. If the unit is divided into shifts, make sure that shift assignments are maintained from day to day, i.e., if a member is assigned to the 0600-1400 shift, keep him on that same shift each day.

Task Allocation and Management

Overloading Service members with tasks or responsibilities is another major source of stress. Allocating tasks fairly among available Service members improves unit effectiveness as well as decreases stress. Proper allocation of tasks include:

Selecting the Right Person for the Job

The right person is fitted to the right task according to the task requirement and the individual's talents, abilities, and training.

Duplicating Critical Tasks

Two Service members are assigned to a critical task requiring mental alertness and complete accuracy. They check each other's work by performing the same task independently.

Cross-Training

Each Service member is trained in a secondary duty position to ensure competently stepping into the position of another.

Developing Performance Supports

Develop standing operating procedures, checklists or other mental aids to simplify critical tasks during periods of low alertness.

Equipment Maintenance and Supply

During pre-deployment, the unit maintains its equipment and manages needed supplies. Once deployed for combat, Service members require confidence that supplies are ample and equipment is dependable. The following questions are important:

- Does the unit provide ample training in equipment maintenance and troubleshooting?
- Has the unit's equipment been field-tested under realistic conditions? For example, have Service members fired and cleaned their weapons while wearing full combat gear or protective clothing?
- Does the unit have sufficient ammunition, food, water, and other essential supplies?
- Does the unit have contingency plans for procuring and managing critical supplies if normal channels are disrupted?

Personal and Family Matters

Family stress adds to combat-imposed stress and causes distraction, interference with performance of essential duties, and a negative impact on stress-coping ability. This will result in the unit's inability to perform at peak. The unit should help the Service members resolve important family care matters before deployment and develop methods for helping families when Service members are deployed. Service members are encouraged to:

- Generate or update their wills.
- Finalize power of attorney for spouses.
- Update life insurance policies, including Servicemen's Group Life Insurance (SGLI).
- Ensure family automobiles are in good repair.

- Develop lists of telephone numbers of reliable points of contact for specific problems (mechanics, emergency transportation, babysitters, sources of emergency money, health care, etc.).
- Resolve major legal issues such as alimony payments, property settlements following divorces, and child support payments.

Role of Leaders

Unit commanders should:

- Brief families as a group before deployment or as soon as possible after deployment. Within the bounds of security limits, explain the mission's nature. Even if a mission is highly confidential, families benefit from such a meeting by being told of the support available to them while separated. They begin to solve problems and form support systems with other families. This includes an opportunity to discuss family questions and concerns. The Family Services Office, base mental health service or the chaplain's office assist in staging this briefing.
- Establish points of contact (e.g. Key Volunteer Network) to assist with family problems. These volunteers possess good working relations with the chaplain and mental health personnel to assist with the management of complex problems.
- Establish key volunteer communication and support networks. Commanding officers' spouses, or spouses of sergeants major are often good resources for developing and running such networks; however, the involvement of junior Service members' spouses is also crucial. Some of the most enthusiastic participants are tasked to make *outreach* visits and encourage shy or depressed spouses to participate.
- Have mental health professionals conduct meetings to discuss post-deployment problems. For example, some children have

difficulty adjusting to a parent's absence, and other family members may experience similar difficulties. Mental health professionals give families valuable information on these normal reactions and suggest ways to prepare for them.

Phase 2: Deployment and Combat

In addition to the normal stress associated with moving to a combat zone, Service members in this phase start worrying about their survival and performance under fire. Their thoughts become centered on fear of the unknown. Leaders should emphasize that stress under these circumstances and conditions is expected and is a natural reaction. This will help prevent "normal" stress reactions from escalating into extreme reactions.

Unit leaders should provide as much information as necessary to their survival and mission success, reinforce the stress control techniques learned during pre-deployment, and help their subordinates understand what happens to them when stressors occur.

Deployment Vehicle

The deployment vehicle—in most cases, an airplane—is a stressor by itself. If it is a commercial aircraft, in-flight problems are usually minor. However, if the unit deploys on a military aircraft, leaders should accomplish the following:

- Designate areas for light exercise and stretching to counter seating discomfort caused by a buildup of leg fluids.
- Ensure Service members drink enough fluids to prevent dehydration and have access to head/latrine.
- Adopt the activity schedule of the new time zone. If the unit is in the sleep cycle or is already in or about to enter the sleep

cycle, cover windows; reduce lighting; and issue earplugs, blankets, and pillows.

- Allow uninterrupted sleep. If a stopover occurs during a sleep cycle, do not waken Service members to eat or partake in activities. If the stopover occurs during an activity period, take full advantage of it by having Service members take washcloth baths, stretch, and perform head-and-shoulder rotations.
- Upon arrival in the area of operations, follow the schedule of the new time zone. Eat the next meal and go to bed on the new schedule. Doing so helps the Service members' bodies adjust.

Information Flow

Since uncertainty about the future is a major source of stress, timely and accurate information becomes vital. Lines of communications are clearly defined and kept open. Issuing warning, operation, and fragmentary orders is critical to ensuring adequate information flow. Informational meetings are conducted at regular intervals, even when there is no new information to disseminate. This reinforces the organizational structure and the importance of unit meetings as the source of current, accurate information. Reliable sources of information are especially important for countering rumors.

Service members also need information or *performance feedback* after mission completion. Merely engaging in a firefight or completing a mission is insufficient. Service members must be told how they performed as a group. The knowledge of mission accomplishment and progress builds unit cohesion, develops a winning attitude, and reduces the effects of stress.

Family Support

Marine Corps Family Team Building programs and corresponding programs for other Services provide family support throughout deployments. The Navy-Marine Corps Relief Society, Army Emergency Relief, American Red Cross, Family Service Centers, and other community agencies also provide direct assistance to family members. The Key Volunteer Network and the American Red Cross continue to function as conduits for emergency information between Service members and their families. Leaders need to educate Service members about these programs and agencies that are available to serve the needs of the community. Effective communication and caring support networks help to prevent anxiety while Service members are deployed and/or in combat.

Religious Ministry Support and Pastoral Care

The Religious Ministry Team is an integrated part of the battalion or unit. In the Marine Corps, a Religious Ministry Team (RMT) consists of a chaplain and Religious Program Specialist (RP) or chaplain's assistant. RMTs are organic to commands or are assigned by higher headquarters to provide direct religious ministry support to a command. (**NOTE:** In the Army, the same teams are called Unit Ministry Teams (UMT).) During operations, the RMT often travels with the forward battalion aid station (BAS) in order to minister to the wounded and to be with the Service members who are most likely to experience combat stress. Using their professional training, skills, knowledge, and relationship with the Service members, chaplains provide care focusing on prevention of mild and moderate combat stress reactions. The RMT's primary mission is to provide ministry and pastoral care to the troops, offering faith, assurance, and hope. Before, during, and after the mission, the team provides field services, sacraments, and counsel, which bring comfort, assurance, and encouragement to Service members as they integrate their experiences into their lives.

In addition to being a spiritual/pastoral mentor for Service members, most chaplains are trained in some form of CED process. Chaplains are effective CED team members as well as trainers of small-unit leaders (e.g., platoon leaders, noncommissioned officers (NCOs), corpsmen, etc.) in CED team member skills and stress management techniques. Appendix B explains the roles of the Religious Ministry Teams.

Physical and Recreational Activities

Service members need an outlet for the anger, frustration, hostility, and grief developed in combat. It is unwise to conduct sports and recreation activities under observation by the enemy, but these activities can be conducted further to the rear. Although makeshift athletic games help relieve stress, they should be conducted with caution. It is interesting to note that the most frequent medical problem in Operation Desert Shield/Desert Storm was sports injury. Other activities, such as listening to music, reading or practicing relaxation exercises are encouraged at rest and recuperation facilities. Time and access to personal hygiene items and facilities are psychologically valuable in combat.

Humanitarian assistance and community relations' projects provide excellent opportunities for Service members to constructively engage in meaningful work. Public Affairs Officers, Civil Affairs Officers, and chaplains are excellent personal resources for developing working relationships with local nongovernmental organizations (NGOs) and private voluntary organizations (PVOs) in the coordination of such projects.

Integration of New Unit Members

The arrival of replacements requires small-unit leaders to conduct continuous training programs. Leaders should be concerned with the various stressors affecting new Service members, as well as those Service members who have been around for a while, including seasoned combat veterans. Leaders teach newcomers to use stress coping skills. Replacements are quickly integrated into their units and become thoroughly acquainted with all aspects of the new unit. The unit also becomes thoroughly acquainted with the replacements. A brief orientation with a sincere welcome, with NCOs and officer support, eases replacement transition for combat.

It is important Service members adjust to a new unit quickly and effectively. If they do not feel they are part of the group, unit cohesion and morale suffer, decreasing unit effectiveness. New Service members in garrison are assigned sponsors to assist them, introduce them to the unit, and smooth their families' adjustment. In combat, replacements are placed with suitable groups of seasoned veterans. New unit members are much more likely to become battle or stress casualties than are members of a seasoned and cohesive group. The veterans need to give support and advice to the newcomers, by example and direct action, if necessary.

Phase 3: Post-Combat

Just as pre-deployment and combat are stressful, the period after combat is also difficult. Today's rapid transportation enables Service members to travel from the battlefield to their hometowns in 48 to 72 hours. This short time often does not give them reflection with their comrades. Units should therefore set aside time in the last few days before leaving the theater to conduct their own "End of Tour" debriefing in which they start at pre-deployment and talk about whatever stands out in their memories, good or bad, as they recount the operation up to its end. There should also be appropriate memorial ceremonies and rituals that formally bring the operation to a close. Awards, decorations, and other recognition must be allotted fairly by the commanders.

Unit officers, staff NCOs, and NCOs, assisted by the chaplains and mental health/CSC teams, prepare the Service members for problems encountered during family reunion. For example, most Service members expect to resume roles and responsibilities they had prior to separation. However, their spouses often resist giving up their new roles as decision makers and primary home managers. Also, a spouse may feel that his or her sacrifices during the Service member's absence have gone unrecognized. This feeling becomes an additional source of tension. If at all possible, the families should receive the same briefings or written materials. Families need to be reassured of their contribution. Key Volunteer Networks and other Marine Corps Family Team building programs, and corresponding organizations for other Services continue to help manage problems with reunion and adjustment.

Service members are briefed that startle reactions to sudden noise or movement, combat dreams and nightmares and occasional problems with sleeping, and feeling bored, frustrated and out of place are common when first returning from combat to a peacetime, civilian setting. The leaders, chaplains, and the CSC team emphasize the normalcy of such reactions. Service members are also advised on resources available to help deal with such symptoms, if they are persistent and become upsetting.

Chapter 3

Command Leadership Actions

3001. LEADERSHIP ACTIONS AND INTERVENTIONS FOR COMBAT STRESS

It is important for military leaders to know something about the treatment of combat stress reactions. As most cases of initial combat stress symptoms do not require medical treatment, it has been found that military leaders are often quite adept at treating less severe cases of combat stress. In most cases, debriefing the mission, including any traumatic events, coupled with rest, food, and sleep, is often sufficient to alleviate the symptoms. If the operational tempo permits the Service member to remain with his unit, and he responds to simple reassurance (e.g., "You just need rest, you'll be okay tomorrow"), he is not a casualty (by definition), and may not require further referral for specialized care. Military leaders in combat often make such decisions. When a Service member requires medical attention to rule out a possible serious physical cause for his symptoms, or because his inability to function endangers himself, the unit, and the mission, he should be transported to the BAS or equivalent nearest medical support facility. Interventions at the small-unit level include:

- If a Service member's behavior endangers the mission, himself, or others, the leader should take appropriate measures to control him.
- If a Service member is upset, let him talk about what is upsetting him, listen, then try to reassure him.
- If a Service member's reliability becomes questionable:

- Unload the Service member's weapon.
- Remove the weapon if there is a serious concern.
- Physically restrain the Service member only when safety is a concern or during transport.

- Reassure unit members that the signs are probably a normal combat stress reaction, and will quickly improve.
- If the combat stress reaction signs continue:
  - Get the Service member to a safer place.
  - Do not leave the Service member alone. Keep someone he knows with him.
  - Notify the senior NCO or officer.
  - Have the Service member examined by medical personel.
- If the tactical situation permits, give the Service member simple tasks to do when not sleeping, eating or resting.
- Assure the Service member that he will return to full duty as soon as possible.

The following quote was taken from an article about fear. It was written by a Marine who articulated his experiences during his tours in Vietnam. It provides intricate insight into how many Service members think and bond during combat.

> What should be done when men succumb to fear brought on by the realization that they are going to die? Before discussing this, first let us look at the men themselves. There are very few men who can be classified as cowards. Most men have too much self-respect to let their buddies down. It is the rare man indeed who will willingly violate the trust of his peers. The vast majority of men will give their lives rather than violate this trust. Proper training of combat troops, prudent leaders who are technically and tactically sound, and the reluctance of men to violate the trust of their peers are the foundation

of a solid combat unit. In this environment, courage and sacrifice are the rule, not the exception. At any given moment, anyone can be rendered ineffective by fear if one realizes that he is going to die. When this happens to a man, do not overreact. He knows he has let his buddies down, and he knows that his buddies and leaders know he has let them down. This is a very uncomfortable feeling for a combat Marine. The men around him will not make a big deal about his actions because they understand the situation. To make an issue of this situation will destroy the man and usually alienate the rest of your men. The man will most likely bounce back to his normal performance. The leader must keep in mind that today's coward is tomorrow's hero. If a man does not bounce back and continues to succumb to this type of fear, the Marines in his fire team will let you know when they have given up on him and no longer consider him trustworthy. Situations such as this are rare, so there is no need to make an example of the man. The biggest pitfall to be avoided is to judge the man's actions as a failure of your leadership as this will indeed cause you to overreact.

3002. SAFETY CONSIDERATIONS

Confiscation of a Service member's weapon should only be considered when it is clearly apparent that the Service member is unreliable and a safety hazard to himself or others. Service members identify with other Service members who are trusted and needed by their comrades. This is one of the strongest factors that aid in returning battle-fatigued members to effective duty. The small-unit leader relocates battle-fatigued Service members to a safer, quieter place to rest and work for several hours up to 1 or 2 days. If the unit cannot wait for the Service members to recover, it is necessary to evacuate them to the first level supporting medical unit. From there, every effort is made to move the Service members to a nonmedical unit or area (a tent or building of opportunity could suffice) for rest, replenishment, and reassurance. It should be made clear that the Service members are tired and in need of an opportunity to talk, to sleep, eat, and replenish fluids; they are not patients. Some small detachments without first sergeants or platoon/company organizations are attached to other units for support and rely on the effectiveness of small-unit leaders. Each Service member is accounted for and every effort is made to ensure strong lines of communications are in place and maintained between Service members and their original unit. Key to successful treatment is the return of the Service member to his original unit.

Actions to be taken for severely combat-stressed Service members are the same as those for the moderately combat-stressed, with one exception: medical personnel at the BAS level should evaluate severely combat-stressed Service members as soon as possible. Casualties will be treated and released within hours; held for rest and replenishment; or evacuated for further restoration. Service members who recover from combat stress reactions; return to their original units, i.e., same company or platoon; and are welcomed upon their return are less likely to suffer recurrence. Once rested and returned, they usually become healthy again. New joins and replacements in the unit are more susceptible to succumb to combat stress. Accordingly, this risk is reduced when Service members recovering from combat stress return to the same unit where their combat experience is known and welcomed. In rare instances, however, it is in the best interest of the individual to be reassigned to other jobs or units.

3003. COMBAT STRESS CASUALTY INTERVENTION MODELS

Management of combat stress reactions is unlike the treatment of physical trauma. Severely injured Service members are stabilized as rapidly as possible and then transferred to the rear. In all wars since World War I, combat stress casualties treated in the rear rarely return to full duty. In contrast, when the same casualties are treated near the front, approximately 75 percent return to full duty. Of those returned to full duty, only 10 percent experience continued symptoms requiring further treatment. Some studies suggest half of those treated at the rear go on to have chronic psychiatric symptoms, and approximately half return to full duty. Therefore, it is clearly in the Service member's best interest to be treated at the front and returned to duty. This also serves the best interest of the unit because it receives a veteran Service member back to duty, rather than breaking in a replacement.

Guidelines for dealing with Service members unable to function because of combat stress reaction are summarized in the memory aid **BICEPS.** BICEPS stands for brevity, immediacy, centrality (Marines)/command contact (Army), expectancy, proximity, and simplicity.

- **Brevity.** Critical Event Debriefing should take 2 to 3 hours. Initial rest and replenishment at medical CSC facilities should last no more than 3 or 4 days. Those requiring further treatment are moved to the next level of care. Since many require no further treatment, military commanders expect their Service members to return to duty rapidly.
- **Immediacy.** CSC should be done as soon as possible when operations permit. Intervention is provided as soon as symptoms appear.

- **Centrality (Marines)/Contact (Army).** Service members requiring observation or care beyond the unit level are evacuated to facilities in close proximity to, but separate from the medical or surgical patients at the BAS, surgical support company in a central location (Marines) or forward support/division support or area support medical companies (Army) nearest the service members' unit. It is best to send Service members who cannot continue their mission and require more extensive respite to a central facility other than a hospital, unless no other alternative is possible. The Service member must be encouraged to continue to think of himself as a warfighter, rather than a *patient* or a *sick person*. The chain of command remains directly involved in the Service member's recovery and return to duty. The CSC team coordinates with the unit's leaders to learn whether the overstressed individual was a good performer prior to the combat stress reaction, or whether he was always a marginal or problem performer whom the team would rather see replaced than returned. Whenever possible, representatives of the unit, or messages from the unit, tell the casualty that he is needed and wanted back. The CSC team coordinates with the unit leaders, through unit medical personnel or chaplains, any special advice on how to assure quick reintegration when the Service member returns to his unit.

- **Expectancy.** The individual is explicitly told that he is reacting normally to extreme stress and is expected to recover and return to full duty in a few hours or days. A military leader is extremely effective in this area of treatment. Of all the things said to a Service member suffering from combat stress, the words of his small-unit leader have the greatest impact due to the positive bonding process that occurs during combat. Simple statements from the small-unit leader to the Service member that he is reacting normally to combat stress and is expected back soon have positive impact. Small-unit leaders should tell Service members that their comrades need and expect them to return. When they do return, the unit treats them as every other Service member and expects them to perform well. Service members suffering and recovering from combat stress disorder are no more likely to become overloaded again than are those who have not yet been overloaded. In fact, they are less likely to become overloaded than inexperienced replacements.

- **Proximity.** In mobile war requiring rapid and frequent movement, treatment of many combat stress cases takes place at various battalion or regimental headquarters or logistical units, on light duty, rather than in medical units, whenever possible. This is a key factor and another area where the small-unit leader helps in the treatment. CSC and follow-up care for combat stress casualties are held as close as possible to and maintain close association with the member's unit, and are an integral part of the entire healing process. A visit from a member of the individual's unit during restoration is very effective in keeping a bond with the organization. A Service member suffering from combat stress reaction is having a crisis, and there are two basic elements to that crisis working in opposite directions. On the one hand, the Service member is driven by a strong desire to seek safety and to get out of an intolerable environment. On the other hand, the Service member does not want to let his comrades down. He wants to return to his unit. If a Service member starts to lose contact with his unit when he enters treatment, the impulse to get out of the war and return to safety takes over. He feels that he has failed his comrades and they have already rejected him as unworthy. The potential is for the Service member to become more and more emotionally invested in keeping his symptoms so he can stay in a safe environment. Much of this is done outside the Service member's conscious awareness, but the result is the same. The more out of touch the Service member is with his unit, the less likely he will recover. He is more likely to develop a chronic psychiatric illness and get evacuated from the war. This is one of the essential principles of CSC.

- **Simplicity.** Treatment is kept very simple. CSC is not therapy. Psychotherapy is not done. The goal is to rapidly restore the Service member's coping skills so that he functions and returns to duty again. Sleep, food, water, hygiene, encouragement, work details, and confidence-restoring talk are often all that is needed to restore a Service member to full operational readiness. This can be done in units in reserve positions, logistical units or at medical companies. Every effort is made to reinforce Service members' identity. They are required to wear their uniforms and to keep their helmets, equipment, chemical protective gear, and flak jackets with them. When possible, they are allowed to keep their weapons after the weapons have been cleared. They may serve on guard duty or as members of a standby quick reaction force.

If treated near their units, 65 to 85 percent of combat stress casualties treated in medical CSC facilities return to duty within 1 to 3 days. About 15 to 20 percent more return to duty in 1 to 2 weeks. Only 5 to 10 percent are sent home, and these usually have other problems in addition to combat stress reactions. If evacuated, few combat stressed Service members will return to duty. In fact, many are likely to be permanently disabled.

History shows that it is important to treat combat stress casualties as close to the front as practicable. Everyone is susceptible to the effects of combat stress, regardless of past performance, rank or duty. *Most Service members receiving combat stress control in accordance with the principles of BICEPS return to duty quickly.* The chances of later psychiatric problems are also reduced.

Combat stress reactions are inevitable in combat, but high stress casualties are not. History shows that highly trained and small cohesive units with good leadership have less than 1 such casualty for every 10 to 15 WIA, even in very heavy fighting. This is significantly fewer than the usual 1:3 to 1:5 ratio in high-intensity battle, and the 1:1 or 1:2 ratio in extreme fighting. Knowledge of factors that increase combat stress disorder in the tactical and overall situation encourages small-unit leaders, comrades, and the individual Service member to act towards building unit cohesion, sharing the burden, and reducing stress. The same measures, plus positive adherence to discipline and the law of war, also prevent combat stress-related misconduct that defeats the purpose of the mission. Tough, realistic training that builds confidence and caring within the unit in combat overcomes many adversities associated with combat stress in a combat environment. However, if a member fails to adequately respond to actions taken by the unit, then he should be evacuated for appropriate medical evaluation.

Chapter 4

Sleep Deprivation

4001. CHALLENGES OF SLEEP DEPRIVATION

People accumulate a "sleep debt" (cumulative loss of sleep over time) when they perform under limited sleep conditions. The only corrective measure for satisfying this sleep debt is sleep itself. Military operations, by their demanding nature, create situations where obtaining needed sleep will be difficult or impossible for more than short periods.

Continuous operations are military operations with many pulses of action every day and night, continuing for several days to weeks, which require careful planning and resource allocation to give everyone a minimum of 4 hours sleep in 24. (FM 22-51)

Sustained operations are continuous operations or combat with opportunity for less than 4 hours sleep per 24 hours for significant personnel, which may be brief or fragmented. (FM 22-51)

Accordingly, service members may have opportunities for only limited or fragmented sleep over an extended period. As a result of these periods of sleep loss, several combat tasks are likely to show decreased performance. These tasks include the following:

- Orientation with friendly and enemy forces (knowledge of the squad's location and maintaining camouflage, cover, and concealment).

- Coordination and information processing (coordinating firing with other vehicles and dismounted elements, reporting vehicle readiness, and communicating with the headquarters).
- Combat activity (firing from bounding vehicle, checking the condition of weapons, observing the terrain for enemy presence).
- Force preservation and regrouping (covering disengaging squads, marking the routes between locations, and conducting reconnaissance).
- Command and control activity (directing location repositioning, directing mounted defense, assigning fire zones and targets).

Continuous operations will potentially be more commonplace on the battlefield. In offensive operations, darkness is the time to retain or gain the initiative; while in defensive operations, obstacles can be employed with greater security during darkness. Forces can disengage undetected and threats to close air support lessen. The physical environment changes at night. As the air cools below ground temperature, inversions reduce visibility and hamper radar and radio signals. Conditions are optimal for using chemical weapons. Visual changes also occur. Without the aid of white light, there is no color perception. There is also a decrease in visual clarity, field of view, and depth perception. Targets take longer to engage. Preparation time increases two-fold to six-fold. Simple actions, such as the departure and return of patrols, become more complex and dangerous. Nighttime planning and coordination require greater attention. Navigation, adjusting fire, and munitions and/or target matching are more difficult. Precision is essential, but accuracy has a price. Service members tend to maintain accuracy at the sacrifice of speed. The adverse conditions associated with or generated by continuous ground combat at night will degrade the fighting performance of Service members, teams, and units. The almost complete mechanization of land combat forces and technological advances that permit effective movement at night, during poor weather conditions, and under conditions of limited visibility have largely overcome the reasons for "traditional" pauses in battle, such as darkness, resupply, and regrouping. New technologies have significantly increased the range, reduced the time, and changed the conditions over which battles are fought. For example, day/night-capable vehicles can operate for extended periods without re-supply, but they are limited by a crew's need to sleep. A Service member is not a machine and is, therefore, the weak link in the chain. The equipment can operate longer than the Service member who operates it, as the Service member must have sleep.

Commanders and leaders must ensure that all Service members obtain enough rest to counteract the effects of rapidly shifting from daytime to nighttime duty hours, or to extended work schedules. Implementing countermeasures that are designed to help Service members adapt to continuous operations conditions can satisfy this requirement. Neither leaders nor their subordinates can perform without rest or sleep. The Service member, the unit, and the leader are all affected by continuous operations. Generally at night, the cognitive and physiological resources of Service members are not at their peak, especially after a rapid shift from daytime to nighttime duty hours. Fatigue, fear, feelings of isolation, and loss of confidence may increase.

Non-stop, unrelieved combat operations (sustained operations) with little or no sleep degrade performance and erode mental abilities more rapidly than physical strength and endurance. Information gained from the Army Unit Resiliency Analysis Model shows that even healthy young Service members who eat and drink properly experience a 25 percent loss in mental performance for each successive 24-hour period without sleep. The mental parameters include decisionmaking, reasoning, memory tasks, and computational tasks. The loss may be greater for Service members who are older, less physically fit, or who do not eat and drink properly.

The effects of sustained operations are sometimes hidden and difficult to detect. Units are obviously impaired when Service members are killed or wounded in action or become noncombatant losses. They are further impaired when their troops are too tired to perform their tasks. Unlike individual performance, unit performance does not deteriorate gradually. Units fail catastrophically, with little warning.

A priority for fighting units is to assure that commanders and leaders are rested and able to think clearly. While this is obvious, it is a most difficult lesson for leaders to learn. During combat, commanders must focus on the human factor. They must assess and strengthen their units as they plan and fight battles. They must accurately decipher which units must lead, which must be replaced, where the effort must be reinforced, and where tenacity or audacity and subsequent success can be exploited. When leaders begin to fail, control and direction become ineffective, and the organization disintegrates. No fighting unit can endure when its primary objectives are no longer coordinated. Leaders must also prepare and precondition Service members to survive. It is particularly important that leaders conscientiously plan and implement effective sleep plans, because activities that are most dependent on reasoning, thinking, problem solving, and decision-making are those that suffer most when sleep and rest are neglected.

Some leaders wrongly believe that their round-the-clock presence during an operation is mandatory; they are unwilling to recognize that they, too, are subject to the effects of sleep deprivation. If the unit has been regularly trained according to the mission command philosophy, two benefits accrue. Not only will a leader be confident that in his absence his subordinates will adhere to his intent, but the trust he shows in his subordinates will continue to maintain unit morale and help ease some of the stress of the situation.

In future operations, the battlefield will become increasingly lethal. The threat of nuclear, biological, and chemical weapons will maximize confusion, uncertainty, and stress, which adversely impact our ability to move, shoot, communicate, and sustain. Sleep loss in this type of environment increases an already stressful situation.

4002. EFFECTS OF SUSTAINED OPERATIONS ON PERFORMANCE

A basic rule for continuous operations is planning ahead to avoid sustained operations, and provide members 5 to 6 hours sleep in 24. However, missions or enemy actions sometimes require exceptional exertion for several days with only unpredictable, fragmented sleep—as required in sustained operations. Sustained combat leads to exhaustion and reduction in effective task performance. Even during the first night of combat, normal sleeping habits and routines are abnormal. The Service member feels the effects of fatigue and the pressure of stress from noise, disrupted sleep time, and threat to life. While essential for endurance, sheer determination cannot offset the mounting effects of adverse conditions. Cognitive degradation involving poor decisionmaking begins during and after the first 24 hours of sleep deprivation.

Individual and unit military effectiveness is dependent upon initiative, motivation, physical strength, endurance, and the ability to think clearly, accurately, and quickly. The longer a Service member goes without sleep, the more his thinking slows and becomes confused. Lapses in attention occur, and speed is sacrificed to maintain accuracy. Continuous work declines more rapidly than intermittent work.

Tasks such as requesting fire, integrating range cards, establishing positions, and coordinating squad tactics become more difficult than well-practiced, routine physical tasks, such as loading magazines and marching. Without sleep, Service members can perform the simpler and/or clearer tasks—lifting, digging, and marching—longer than the more complicated or ambiguous tasks such as a fine hand-eye coordination sequence; i.e., tracking a target through a scope.

Sleep loss affects memory, reasoning, mental assessments, decision-making, problem-solving, subsequent actions, and overall effectiveness. While comprehension is accurate, reading speed slows and recall fails. For example, Service members may understand orders when reading them in documents, yet they are forgotten later when required. Individuals will forget or omit assigned tasks more often than they will make errors in carrying them out.

Leaders can expect declining moods, motivation, initiative, planning ability, and preventive maintenance. High motivation will only increase risk, due to impaired performance. Leaders must recognize erratic or unreliable task performance in subordinates, as well as in themselves. Alertness and performance decline gradually with partial sleep deprivation; that is, when sleep is limited to 4 to 5 hours per night. After 5 to 7 days of partial sleep deprivation, alertness and performance decline to the same low levels as those following 2 days of total sleep deprivation. After 48 to 72 hours without sleep, personnel become militarily ineffective.

Adverse Conditions

Continuous combat forces Service members to perform under adverse conditions that cause degradation in performance. Examples of adverse conditions follow.

Low Light Level

The amount of light available for seeing landmarks, targets, and maps is greatly reduced at twilight and night.

Limited Visibility

Smoke, fog, rain, snow, ice, and glare degrade a Service member's ability to see his environment and objects within it, as opposed to situations free of such conditions.

Disrupted Sleep Routines

People are accustomed to being awake or asleep during certain hours of the day or night. Disruption of the normal sleeping schedule causes degraded performance.

Physical Fatigue

Working the muscles faster than they can be supplied with oxygen and fuel rapidly creates "oxygen debt," eventually making these muscles unable to function until the deficits are made up during brief rests.

Sleep Loss

The muscles can continue to function adequately without sleep, but the brain cannot. Increasing sleep debt leads to subtle, but potentially critical, performance failures.

Sleep Loss Indicators

Indications of degraded performance symptoms become more prevalent as sleep debt accumulates. Performance is affected by the hours of wakefulness, tolerance to sleep loss, and the types of mental or physical work. Both mental and physical changes occur, with symptoms varying among individuals. Leaders must observe Service members for the following indications of sleep loss and degraded performance:

- Physical changes in appearance, including vacant stares, bloodshot eyes, pale skin, and poor personal hygiene. Other physical signs of sleep loss include the body swaying when standing, sudden dropping of the chin when sitting, occasional loss of hand-grip strength, walking into obstacles or ditches, low body temperature, slowed heart rate, and slurred speech.
- Mood changes, decreased willingness to work, and diminished performance go hand-in-hand. Service members may experience decreasing levels of energy, alertness, interest in their surroundings, and cheerfulness with a concurrent increase in irritability, negativity, and sleepiness. Some become depressed and apathetic. Others, for a time, can become energized by sleep loss, talk more, and may be more assertive without necessarily maintaining good judgment. Sleepiness and mood changes are not signs of weakness. After long periods of sleep loss, Service members go from being irritable and negative to dull and weary.
- Service members may feel more effort is needed to perform a physical task in the morning than in the afternoon. Exaggerated feelings of physical exertion may lead to work stoppage, especially between 0400 and 0700. During that time, the tendency to fall asleep is considerably more noticeable than other times.
- Both bickering and irritability increase with sleep loss. When Service members argue, it shows that they are still talking to each other and exchanging orders and messages. When arguments cease, especially after a period of increased bickering, Service members may be in a state of mental exhaustion.
- Comprehension and perception slow considerably. Individuals require extended time to understand oral, written or coded information; to find a location on a map and/or chart coordinates; to interpret changes in enemy fire patterns; and to make sense of things seen or heard, especially patterns. They may

have difficulty with spot status or damage reports, and may be unable to assess simple tactical situations.

Loss of Concentration

Sleep deprivation causes the attention span to shorten. There is a loss of concentration on the job as dream-like thoughts cause lapses in attention. Leaders should watch for the following:

- Decreased vigilance. Personnel are less alert and fail to detect the appearance of targets, especially in monotonous environments. They may doze off at the wheel of moving vehicles.

- Distorted attention. Service members may imagine seeing things that are not there, e.g., "moving" bushes when in reality there is no such movement. The sleep-deprived brain can also misperceive bushes, rocks, people, vehicles or anything else and see them as something different, in very precise detail. Often the tired brain "sees" what it wishes were there (food, a bed); at other times, these illusions may be animals or other more bizarre things. But when the mind is alert for an enemy, the brain may generate a very convincing, detailed image of the enemy. Sometimes, but not usually, sounds or other sensations may accompany these illusions. They usually last only seconds, but can persist for minutes if not challenged, and rarely have even been "seen" by equally sleep-deprived comrades when told of them. It is essential for sleep-deprived unit members to check out any questionable things they see with their comrades, and to faithfully follow reporting and challenge procedures.

- Inability to concentrate; easily confused. Service members cannot keep their minds on what they are doing. They cannot follow multiple directions nor perform numerical calculations.

- Failure to complete routine tasks. Sleep loss interferes with completing routine individual tasks, such as drying the feet, changing socks or filling canteens when water is available. Tasks such as performing weapons checks may be skipped.

When a Service member cannot recall what he just saw, read, heard or was told by another individual, he is exhibiting a common sign of sleep loss. His memory loss is limited to recent events. For example, a sleep-deprived Service member may forget recent target data elements or recall them incorrectly and have difficulty learning new information.

4003. ACHIEVING SLEEP IN COMBAT

Sleep deprivation produces stress and, therefore, sleep management is important. Sleep management is a combat multiplier. Planned sleep routines are important for keeping the unit, the individual Service members, and the leader himself functioning as required while reducing sleepiness during continuous combat. Since leaders are responsible for planning sleep routines, they need a basic understanding of the physiological and behavioral aspects of sleep and their impact on performance. The following paragraphs provide this information.

Rhythmic Variations

There are rhythmic variations in individual performance based on a predictable physiological and behavioral cycle that comprises about 24 hours. The 24-hour, day-night/work-rest cycle is called the *circadian rhythm*. Because traveling across a half-dozen time zones disrupts the usual relationship in the day-night/work-rest cycle, for a few days Service members are not sleepiest at their usual sleep period of 2400 to 0600, new-locale time. Allowing sleep about 1200 to 1800, new-locale time, will only delay their adaptation to their new locale. Leaders must instruct troops to go to bed between 2400 and 0600 new-local time to establish a new circadian rhythm.

Another example of circadian rhythm is body temperature. Although one's "normal" temperature is 98.6 degrees, this is really an average or midpoint of a daily swing from 96.8 to 100.8 degrees. For someone accustomed to working days and sleeping nights, body temperature would fluctuate approximately as indicated. There is a well-established link between body temperature and sleepiness and/or performance slumps. Performance parallels body temperature. The higher the body temperature, the better the performance. As body temperature decreases, mood and motivation decline with a concurrent increase in sleepiness and fatigue.

Impact upon performance is most pronounced during the *circadian lull*, which is roughly 0200 to 0600 hours. During this time, performance declines about 10 to 15 percent. In sleep-deprived Service members, this decline may reach 35 to 40 percent. If the day-night/work-rest cycle is disrupted, performance suffers because the Service member is sleepy during the new work period and awake during the new sleep period. The body needs several days to adjust to the new schedule. Critical hours for sleep are between 0200 and 0600 when *anchor sleep* (the most beneficial sleep) is taken. The body is at its lowest temperature during this period. This is the best time for sleeping, but not for napping. To prevent sleep inertia, naps should always be taken at times other than the lowest point in body temperature.

Leaders need to calculate the difference in time zones and make the necessary schedule changes. Leaders will need day-and night-fighting teams. Members acclimated to working days and sleeping nights should be scheduled to work nights and sleep days.

Their performance slump/optimal time to sleep would be 2400 to 0600, new-locale time. Deployment, pre-combat, and combat are not usual circumstances. If certain Service members must have an offset circadian timing from the rest of the unit, a special effort must be made to establish their sleeping time. Obviously, troops must sleep whenever possible. If a planned sleep schedule cannot be followed, however, performance is enhanced if sleep coincides with the low point in body temperature.

Adjusting to new circadian rhythms is a slow process, taking 3 to 6 days to come "in phase" with a new schedule. Leaders should devise a sleep schedule that provides for sleep at the same time of day or night every 24 hours. Sleep schedules that provide for sleep at different times of day or night are less valuable and are detrimental to quality sleep and optimal performance.

Sleep Shifts

Staggered work schedules can be set up for two shifts working 4 hours on/4 hours off, 6 hours on/6 hours off, and 12 hours on/12 hours off. See Table 4-1. Each shift follows the same schedule daily. It is better to maintain regular shift schedules than schedules that continually change.

Sleep/Rest Guidelines

Leaders should use the following sleep and/or rest guidelines in this section to enhance individual and the unit performance in continuous operations.

- Know personal tolerance for sleep loss and those under your command; major individual differences are not easily changed. Individuals who are unable to sleep during pre-deployment and deployment stages should be encouraged to practice relaxation exercises (see paragraph 2005).

Table 4-1. Sleep Shifts.

4 HOURS ON/4 HOURS OFF						
Shift	2400-0400	0400-0800	0800-1200	1200-1600	1600-2000	2000-2400
1	SLEEP	DUTY	SLEEP	DUTY	SLEEP	DUTY
2	DUTY	SLEEP	DUTY	SLEEP	DUTY	SLEEP

6 HOURS ON/6 HOURS OFF				
Shift	2400-0600	0600-1200	1200-1800	1800-2400
1	SLEEP	DUTY	SLEEP	DUTY
2	DUTY	SLEEP	DUTY	SLEEP

12 HOURS ON/ 12 HOURS OFF		
Shift	2400-1200	1200-2400
1	SLEEP	DUTY
2	DUTY	SLEEP

- Ensure that Service members fully use their breaks and other opportunities for rest. Encourage them to waste no time in getting to sleep. Undisturbed, prolonged sleep is the most desirable use of rest opportunities. When there has been sleep loss but little physical exertion (e.g., manning communications, operating a radio), mild physical exercise such as walking around when conditions permit, can help maintain alertness.
- Encourage Service members to sleep, not just rest, by creating the most conducive environment possible for sleep: quiet, without interruptions (or earplugs); dimness or darkness (or with eye cover); not overly warm or cold.

- Do not allow personnel to sleep in unsafe conditions. Enforce strict rules designating sleep areas and requiring perimeter guards. Require day and night guides for all vehicles to prevent Service members from being accidentally run over.
- Ensure that Service members follow sleep schedules or routines. The field commander who does not enforce a sleep schedule or routine leads his troops into an environment that increases the opportunity for hazardous conditions to be encountered while in continuous combat. Taking naps is not a sign of low fighting spirit or weakness; it is a sign of foresight.

Measuring Sleep Loss

Sleep loss can be measured by:

- Keeping a sleep and/or activity log. From pre-deployment to post-deployment, log sleep and nap periods. Service members need 4 to 5 hours per 24-hour period; 6 or 7 hours is optimum. If they receive less, the first chance for a long rest period must be used for sleep.
- Observing performance and asking questions. Look for the indications of sleep loss—such as increase in error occurrence, irritability, difficulty understanding information, and attention lapses—with concurrent decreases in initiative, short-term memory, and attention to personal hygiene. Confirm sleep loss by asking the obvious question: "When did you sleep last and how long did you sleep?"

Sleep Loss Alternatives

Ways to overcome performance degradation include:

- Upon signs of diminished performance, find time for members to nap, change routines or rotate jobs (if cross-trained).

- Have the Service members most affected by sleep loss execute a self-paced task.
- Have Service members execute a task as a team, using the buddy system.
- Do not allow Service members to be awakened for meals while in flight to a new location, especially if the time zone of the destination is several hours different than that of point of departure.
- Insist that Service members empty their bladder before going to bed. Awakening to urinate interrupts sleep, and getting in and out of bed may disturb others and interrupt their sleep.
- Allocate sleep by priority. Leaders, on whose decisions mission success and unit survival depend, must get the highest priority and largest allocation of sleep. Second priority is given to Service members that have guard duty and to those whose jobs require them to perform calculations, make judgments, sustain attention, evaluate information, and perform tasks that require a degree of precision and alertness.

4004. SLEEP/REST PLANNING

Sleep/rest planning applies to the pre-deployment, deployment, pre-combat, combat, and post-combat stages of battle.

Pre-Deployment Stage

Using mission-scenario operation guidelines, determine periods available for sleep and the total number of sleep hours possible. Because continuous operations requirements may change, alternate sleep routines should be planned. Become familiar with the area where the combat unit will sleep; For example, some may have to sleep in mission-oriented protective posture (MOPP) IV. If sleeping in MOPP IV is anticipated in combat, practice it during the pre-deployment stage. Prior experience reduces stress, so practice anticipated sleep routines before continuous operations.

Deployment Stage

Since sleep will be reduced during deployment, follow preplanned sleep routines. The prudent commander will choose a 4-hour on/4-hour off, 6-hour on/6-hour off, or 12-hour on/12-hour off shifts from the start. Take into account that Service members on night duty will need to sleep during the daytime. Provide night-shift personnel with separate sleeping quarters to avoid disruption of their sleep period.

Pre-Combat Stage

In general, people are most effective during the afternoon and are least effective just before dawn. Without prior adjustment to the new time zone, which naturally occurs in 3 to 5 days, leaders can expect degraded daytime performance. The reason is that 0200 to 0600 hours home-base time is the low point in performance efficiency and should be considered when planning workloads.

Combat Stage

Every effort should be made to avoid situations where all personnel are physically and mentally exhausted simultaneously. Make the most of any lull during the combat phase by sleeping briefly. Complete recovery from sleep loss may not be possible during intense combat, but limited sleep is helpful. Uninterrupted short sleeps of 15 minutes or longer are beneficial to partially recovering alertness. Sleep during the combat stage may be risky, however, because a Service member may wake up feeling groggy, confused, sluggish, and uncoordinated. It may take his brain from several seconds to 15 minutes to "warm up." Individuals differ in how quickly they take to wake up, but it tends to be worse when the body expected to go into deep sleep, and to get worse with increasing sleep loss. Activities that increase circulation of warm blood to the brain, like moderate exercise or drinking a hot beverage, may shorten the start-up time.

Post-Combat Stage

It is important to make up sleep debt, but experts disagree about the amount of recovery time needed. Some say the hours of sleep needed for recovery after sleep deprivation are less than the amount lost. It is well known and documented that lost sleep is not made up hour-for-hour. Most experts agree that immediately following continuous combat, Service members should be allowed to sleep up to 10 hours. Longer sleep periods are not desirable because they cause "sleep drunkenness" and delay in getting back to a normal schedule. After the first sleep period of up to 10 hours, Service members should return to the regular sleep routine. Sleep inertia lasting longer than 5 to 15 minutes and increased sleepiness may occur for as long as a week following sustained combat. Some experts recommend that 4 of the first 8 hours of recovery sleep should be at the 0200 to 0600 sleep time, and they suggest the following guidelines for complete recovery from the effects of sleep loss:

- 12 hours for sleep and rest after 36 to 48 hours of complete sleep loss with light to moderate work load (fatigue may linger for 3 days).
- 24 hours for sleep and rest after 36 to 48 hours of sleep loss with high workload (12 to 16 hours per day).

- 2 to 3 days time off after 72 hours or more of acute sleep loss.
- As much as 5 days for sleep and rest following 96 hours or more of complete sleep loss.

Most experts agree that 10 hours of sleep is the maximum needed, with the additional 2 hours used for rest. It is doubtful that a Service member could continue past 72 hours of wakefulness. Should this occur, a couple of nights with 10 hours of sleep are more beneficial than an excess of 10 hours during one sleep period. If Service members have not slept for 36 to 48 hours or more, they should avoid sleep of less than 2 hours, especially between 0400 and 0600. A too-short sleep period at the wrong time may cause a long period of sleep inertia. After 96 hours of total wakefulness, 4 hours of sleep may provide substantial recovery for the simpler, less-vulnerable tasks. Recovery continues with additional days of 4 hours of sleep per 24 hours. Complex leadership tasks may require longer recovery sleep, but sleep until fully satisfied is not necessary.

Sleep loss alone does not cause permanent health problems, nor does it cause mentally healthy people to become mentally ill. Reduced sleep (from 8 to 4 hours) does not cause physical harm. Hallucinations may occur, but they disappear after recovery sleep. Clinical laboratory tests show that total sleep loss of over a week does not pose serious health problems. It is doubtful that Service members could stay awake for such an extended period, and it is not suggested that Service members try to endure long periods without rest. However, the effects of sleep loss, such as inattentiveness and poor judgment, may be harmful (such as falling asleep at the wheel of a vehicle).

Sleep cannot be stored in our bodies for emergency use. Sleep of more than 7 to 8 hours before deployment does not "store up" excess sleep, but sleep taken immediately before a deployment can prolong activity. Therefore, it is important to begin continuous operations in a rested state. During daytime or early morning naps, many Service members experience vivid dreams as they fall asleep and often wake up frightened. Leaders should inform their troops that this occurrence is both common and normal during daytime sleep. If a single, unbroken period of 4 to 5 hours is not available for sleep, "power naps" of 15 to 30 minutes, although less recuperative, can be taken. Leaders must capitalize on every opportunity for a "power nap." Merely resting by stretching out does not take the place of sleep. Only sleep can satisfy the need for sleep.

Appendix A

Suicide Awareness Information

SIGNS OF SUICIDE

A person contemplating suicide—

- Believes he or she is in a hopeless situation.
- Appears depressed, sad, tearful; may have changes in patterns of sleep and/or appetite.
- May talk about or actually threaten suicide, or may talk about death and dying in a way that strikes the listener as odd.
- May show changes in behavior, appearance or mood.
- May increase or start drug or alcohol use.
- May injure self or engage in risky behavior.
- Abandons planning for the future.
- May start withdrawing from others, including family and close friends.
- May give away possessions.
- May appear apathetic, unmotivated, indifferent.

PREVENTION KEYS

The small-unit leader may be the first to identify the "signals" of a potential suicide-risk Service member. The keys to prevention are to provide aid to persons at suspected risk and follow the acronym AID LIFE.

A	Ask: "Are you thinking about hurting yourself?"
I	Intervene immediately.
D	Do not keep a secret concerning a person at risk.
L	Locate help (NCO, chaplain, corpsman, doctor, nurse, friend, family, crisis line, hospital emergency room).
I	Inform your chain of command of the situation.
F	Find someone to say with the person. Do not leave the person alone.
E	Expedite! Get help immediately. A suicidal person needs the immediate attention of helpers.

Appendix B

Religious Ministry Team's Role

(Adapted from Army Field Manual (FM) 22-51, *A Leader's Manual for Combat Stress Control.*)

This appendix addresses the general role of the ministry team in the commander's program of combat stress control and in battle fatigue ministry. In the Marine Corps ministry teams are called Religious Ministry Teams (RMTs) and consist of chaplains and Religious Program Specialists (RPs) or chaplain's assistants (In the Army they are called Unit Ministry Teams). The RMT is assigned to a command or designated by higher headquarters to be responsible for the direct religious ministry support to the command. RMTs provide professional ministry support to leaders in fulfilling their combat stress identification and intervention responsibilities. RMTs can also assist in training leaders to recognize combat stress symptoms.

EFFECTS OF STRESS

Negative effects of stress can be lessened when members are prepared physically, emotionally, and spiritually prior to combat/operations. RMTs can assist in preparing Service members to manage combat stress with training before and during deployment. This training helps Service members draw upon their personal religious faith, spiritual strength, and values as well as to share strength and confidence during any operation.

RELIGIOUS MINISTRY TEAM RELATIONSHIP

The RMT's established relationship with the command promotes trust with the troops. As integral members of their commands, RMTs are trained and ready to respond to the needs of Service members experiencing combat stress. A person-oriented resource, the chaplain and RP team minister to combat stress casualties, especially Service members having less severe difficulties who have rapid replenishment potential.

SPIRITUAL VALUES

Service members' inner resources are often based on their personal faith, religious beliefs and spiritual values. In combat, Service members may show more interest in their religious beliefs. When religious and spiritual values are challenged during the chaos of combat, Service members may lose sight of inner resources that sustain them. Service members can then become targets of fear, despair, hopelessness, and eventually, combat stress casualties. They are also at risk for committing misconduct stress behaviors. Chaplains are a source of direction and stability to Service members experiencing these dilemmas and seeking to refocus their personal beliefs and spiritual values.

RELIGIOUS MINISTRY TEAM SUPPORT

Ministry teams can provide preventive, immediate, and replenishing spiritual and emotional support and care to Service members experiencing combat stress.

Preventive

The RMT's religious ministry support assists in preventing combat and misconduct stress behaviors through establishment of a presence within the unit. It is important for the RMT to be present with Service members when the unit trains and when it deploys. The chaplain and RP can be a calming influence on members and can help Service members strengthen or regain values important to them. Chaplains can help prevent combat stress and misconduct by:

- Being present with the Service members and deploying with the unit.
- Providing opportunities for private and group prayer and worship.
- Supplying personal religious articles and materials.
- Reading the scriptures with Service members.
- Providing sacraments as the situation allows.
- Counseling Service members, allowing them to work through stress, fear, anxiety, anger, and frustration.
- Visiting Service members in work and living areas.
- Assisting Service members and families prior to deployment with preparation for geographical separation and an uncertain future through Marine Corps Family Team Building (MCFTB) programs that emphasize family strengths. This helps Service members to know that their families are cared for during deployment.

Immediate

RMTs can assist commanders in the identification of Service members experiencing combat stress. Chaplains often work closely with the medical officer and are trained to recognize the signs of combat stress and provide religious support to Service members experiencing combat stress. Chaplains assess spiritual needs and provide the appropriate religious ministry. This may include such things as:

- Presence with the Service member.
- Conversation and counseling, providing opportunities to share fears, hopes, and other thoughts.
- General prayers; prayers for the individual or fallen comrades.
- Rites, sacraments, and ordinances as appropriate.
- Reading from scriptures.

Replenishing

Following an engagement, the unit may require reconstitution through the addition of new personnel. The RMT can assist surviving Service members rebuild their emotional, psychological, and spiritual strength. During this time, RMTs continue their ongoing direct religious ministry that includes:

- Coordinating the availability of worship services, sacraments, rites, and services and/or ceremonies honoring the dead.
- Assisting with the integration of personnel replacements.
- Enabling the grief process through personal counseling and memorial services.

- Reinforcing the Service member's sense of self-worth and hope.
- Assisting with CEDs to provide opportunities for Service members to talk about their combat experiences and to facilitate integration of the combat experience into their lives.
- Providing leadership training and supervision of CEDs.
- Providing personal religious articles and materials.
- Providing or arranging faith group coverage in the command or for other units.
- A chaplain's ability to relate the spiritual aspects of life to the Service member's situation is an essential element in the replenishment process. Religious ministry can assist Service members in achieving emotional and spiritual wholeness.

Reintegration

After deployments and operations, RMTs can assist reintegration of the individual Service member into family relationships and society at large. Many religious ministry support programs remain the same. Expanded ministry may also include providing the following:

- Worship events for the entire unit.
- Worship events for varied faith groups.
- Briefings that help Service members recognize, prepare for, and master the stressors of reunion with family.
- Structured events to assist Service members returning to family and civilian life.
- Opportunities for Service members to experience and understand the forgiving and unchanging love of God.

Appendix C

Glossary

Section I. Acronyms

AAR .... after action review
BAS ....battalion aid station
BICEPS .... brevity, immediacy, centrality, expectancy, proximity, and simplicity
CED ....critical event debrief
CSC .... combat stress control
MCFTB ....Marine Corps Family Team Building
MCRP..................................Marine Corps Reference Publication
MOPP.................................. mission-oriented protective posture
MRE .... meals, ready to eat
NCO .... noncommissioned officer
NGO .... nongovernmental organization
PVO..private voluntary organization
RMT .... religious ministry team
RP..religious program specialist
SGLI....................................Servicemen's Group Life Insurance
SPRINT .... Special Psychiatric Rapid Intervention Team
UMT...unit ministry team
WIA...wounded in action

Section II. Definitions

circadian rhythm—Rhythmic variations in individuals; predictable physiological and behavioral cycle of about 24 hours.

combat stress—The mental, emotional, or physical tension, strain, or distress resulting from exposure to combat-related conditions. Combat stress is the net result of many stressors (e.g., fear of death, fear of failure, other intensive painful emotions [like grief], uncertainty, boredom, worry about what is happening back home, and the many physical and mental demands of combat duties) at any given time.

combat stress behaviors—The generic term that covers the entire range, from highly positive to totally negative behaviors which can be elicited by combat stress.

combat stress reaction—The common, predictable, negative, emotional and physical reactions of normal people to the "abnormally high" stressors of combat, which by definition interfere with mission performance or well-being, but which can be treated according to the BICEPS principles, with reassurance of normality and recovery, with rest, replenishment, and activities to restore confidence. The combat stress reactions do not include the misconduct stress behaviors, since the latter require disciplinary action.

continuous operations—military operations with many pulses of action every day and night, continuing for several days to weeks, which require careful planning and resource allocation to give everyone a minimum of 4 hours sleep in 24.

stress—An internal process of an individual for preparation in dealing with, or reacting to stressors. Stress involves physiological stress reflexes, which ready the body for fight or flight, such as increased release of adrenaline, heart rate, and blood flow to brain and muscles, with subsequent increase in alertness. Stress also involves the accompanying emotions (stress responses) such as anxiety, anger, or guilt. Stress may or may not involve conscious awareness of the threat, but the stressor must be perceived at some level of awareness. The amount of stress experienced depends on the appraisal of the stressor and its context by the individual, even if that perception is incorrect. The stress process includes psychological defenses which may shield the individual from perceiving more threat than he can tolerate.

stress behaviors—Stress behaviors are the observable behaviors that an individual exhibits as a result of internal stress. They may be intended to overcome and turn off the stressor, escape it, adapt to an inescapable stressor, or make the uncomfortable stress symptoms more tolerable. A stress behavior may simply be a side effect of the tension, which is part of stress. Any of these different types of stress reactions may be successful, unsuccessful, adaptive, or maladaptive. They may make the situation better, worse, or not influence it at all. They may resolve one stressor but create new ones.

stressor—Any event or situation that requires a non-routine adaptation or behavior change, creates conflicting motives within the individual, or is perceived by the individual as a threat to his well-being or self-esteem. Combat stressors are any that occur in the context of performing the combat mission (whether under fire or not).

sustained operations—Continuous operations or combat with opportunity for less than 4 hours sleep per 24 hours for significant personnel, which may be brief or fragmented.

Appendix D

References

DoD Directive

6490.5, *Combat Stress Control (CSC) Programs*

Army Field Manual (FM)

8-51, *Combat Stress Control in a Theater of Operations, Tactics, Techniques, and Procedures*

22-9, *Soldier Performance in Continuous Operations*

22-51, *Leaders' Manual for Combat Stress Control*

22-100, Army *Leadership*

90-22, *(Night) Multiservice Night and Adverse Weather Combat Operations*

Miscellaneous

U.S. Army Aeromedical Research Laboratory and U.S. Army Safety Center, *Leader's Guide to Crew Endurance*

http://www.mcu.usmc.mil/TbsNew/Coverpag.htm, *Human Factors in Combat: Topics in Combat Leadership that Help Preserve the Fighting Spirit and Prevent Psychiatric Casualties*

*TC 90-1

Training Circular
No. 90-1

Headquarters
Department of the Army
Washington, DC, 19 May 2008

Training for Urban Operations

Contents

Page

Distribution Restriction: Approved for public release; distribution is unlimited.

*This publication supersedes TC 90-1, 1 April 2002.

Figures

Tables

Chapter 1

Introduction

This chapter discusses definitions, training strategy, prerequisite training, individual task training, and collective task training for urban operations (UO). Military operations on urbanized terrain (MOUT) and UO differ in ways that reflect the changes in the way we engage in war today.

Section I. DEFINITIONS

This section defines and compares MOUT and UO.

Note: The terms "urban operations" and "UO" replaces all instances of "military operations on urbanized terrain" and "MOUT."

MILITARY OPERATIONS ON URBANIZED TERRAIN

1-1. In the past, Army doctrine characterized urban areas as a type of specialized terrain, like jungles and mountains, and called operations in this type of terrain *military operations on urbanized terrain* (MOUT). Doctrine emphasized tactical warfighting tasks on this terrain, which was characterized by densely situated buildings and streets, and assumed the enemy consisted of conventional (Soviet-type) forces. Defeating or destroying those forces overrode all other considerations.

URBAN OPERATIONS

1-2. The concept of UO assumes that the urbanized terrain is populated, and that the populace must be a foremost consideration. UO recognizes that the populace can help either side, and that it requires a working infrastructure. Urban operations, whether against an insurgent or unconventional enemy, thus vastly differ from MOUT (Table 1-1).

Table 1-1. Comparison of MOUT and UO.

	MOUT	*UO*
Enemy force	Conventional, Soviet type	Insurgent or unconventional
Mission	Defeat or destroy enemy above all other considerations	Defeat the enemy Gain and retain the good will of the populace
Populace	Not considered	Important consideration

1-3. Although certain techniques and tactics are common to both types of operations, UO goes beyond MOUT, focusing not only on defeating the enemy, but also on gaining and retaining the good will of the population. UO consider the complex and dynamic interactions and relationships between the urban area's key components--the terrain (natural and manmade), the population, and the supporting infrastructure--as an overlapping and interdependent system of systems.

1-4. A single urban operation may include the full range of Army operations--offensive, defensive, stability and support (Appendix C)--that may be executed, either sequentially or simultaneously (FM 3-06, Chapter 1). Furthermore, to accomplish an objective, UO demand working with sister services, US government agencies, nongovernmental organizations, and regional and international organizations.

Section II. TRAINING STRATEGY

The UO training strategy is primarily organized by environment and phase. Training environments include live, virtual, and constructive (Figure 1-1), and are based on the unit's METL. Training phases are initial, intermediate, and culminating. Table 1-2 shows the relationship between the training environments and the first three training phases. (See also Appendix D, which discusses leader training.)

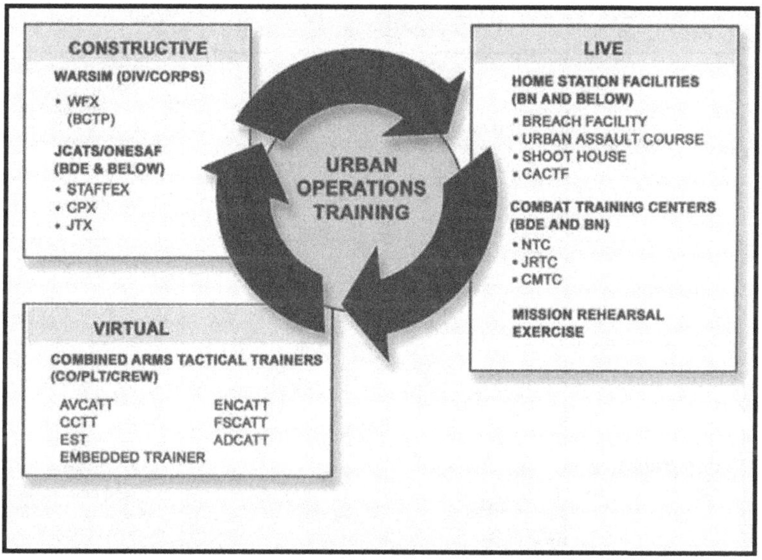

Figure 1-1. Urban Operations Training Support Strategy.

TRAINING PHASES

1-5. The three training phases are initial, intermediate, and culminating. Each of these includes some live, virtual, and constructive training.

INITIAL PHASE

1-6. This phase, which trains individual and collective tasks, is the first step in the building-block approach. It forms the basis of unit or collective training. For Soldiers and leaders, mastery of individual and crew skills is a prerequisite for UO training. The tasks train basic marksmanship skills and other Soldier tasks needed to perform any tactical operation. Included are training in a live environment (precision marksmanship) and virtual environment (Engagement Skills Trainer [EST], crew simulators, and CCTTs).

Table 1-2. Training environment/training phase matrix.

TRAINING PHASE / TRAINING ENVIRONMENT	INITIAL	INTERMED	CULMINATING
LIVE	Precision Marksmanship (Indiv / Sqd)	Breach Facility, Urban Assault Course, Shoot House (Team / Squad / Platoon)	CACTF, Platoon STX, Company STX, Battalion FTX Deployment CTC Rotation Mission Rehearsal Exercise
VIRTUAL	EST (Indiv / Sqd) Crew Simulators and CCTT (Crews Plt Ldrs, Co Cdrs)		
CONSTRUCTIVE			WARSIM (division/corps), which includes warfighting exercise (battle command training program). JCATS/OneSAF (brigade and below), staff exercises (STAFFEXs), command post exercises (CPXs), and joint training exercises (JTX)

INTERMEDIATE PHASE

1-7. This phase of training is characterized by collective task training in varied conditions such as limited visibility; different entry techniques and points; incorporation of combatants, noncombatants, and OPFOR; use of MILES; and live-fire exercises. It is the vital link between individual tasks and mission accomplishment. It should be conducted at, but is not limited to, the *Breach Facility, Live-Fire Shoot House*, and *UAC*. (Section III discusses these facilities in detail.) Many types of organizations can use these facilities. Trainers *must* be certified before conducting demolition training IAW local standing operating procedure (SOP) or ACOM policy. AR 385-63 prescribes safety requirements. Except for Aviation, Armor, and other branches that have no practical use for these facilities, this phase must be completed to advance to the culminating phase.

CULMINATING PHASE

1-8. The Combined Arms Collective Training Facility (CACTF, Figure 1-2) is where the culminating phase of urban operations training ends. This phase can include multiechelon, combined arms, and branch specific STX training may be conducted. Urban operations leader training occurs in leader professional development courses (Appendix D). Units have developed internal urban warfare training programs and TRADOC schools have assembled technical mobile training teams (MTTs) for marksmanship, combatives, and breaching techniques.

EDUCATION PHASE

- Urban warfare reading/video list.
- Urban operations equipment.
- Review of Infantry doctrinal publications (FM 3-21.10, FM 3-06.11, for Infantry), and corresponding publications for other branches.
- Small-unit urban movement techniques and considerations.

PLANNING AND EXECUTION PHASE

- Prepare company OPORD or WFF-related products.
- Brief platoon/company OPORD/WFF-related products to commander or staff representative.
- Participate as briefer in urban environment TEWT.

SUSTAINMENT PHASE

- Participate in company/platoon TEWT or other urban operations event.
- Execute/facilitate a company or platoon TEWT.

TRAINING ENVIRONMENTS

1-9. Both live and virtual training environments are used to train UO.

LIVE ENVIRONMENT

1-10. The live training environment consists of home station training using the Urban Assault Course (UAC), Shoot House (SH), Breach Facility (BF), Combined Arms Collective Training Facility (CACTF), Combat Training Center (CTC) rotations, and any other live training that supports the UO training.

VIRTUAL ENVIRONMENT

1-11. The virtual training environment consists of simulators such as Close Combat Tactical Trainer (CCTT) or Engagement Skills Trainer (EST), which are used to train small-unit leaders and crews. The constructive environment consists of battle staff training for battalion and above staff and leaders such as Joint Conflict and Tactical Simulation (JCATS)/OneSAF.

CONSTRUCTIVE ENVIRONMENT

1-12. The constructive training environment consists of the following, which all occur in the culminating training phase:

- WARSIM (division/corps), which includes warfighting exercise (battle command training program).
- JCATS/OneSAF (brigade and below), staff exercises (STAFFEXs), command post exercises (CPXs), and joint training exercises (JTX)

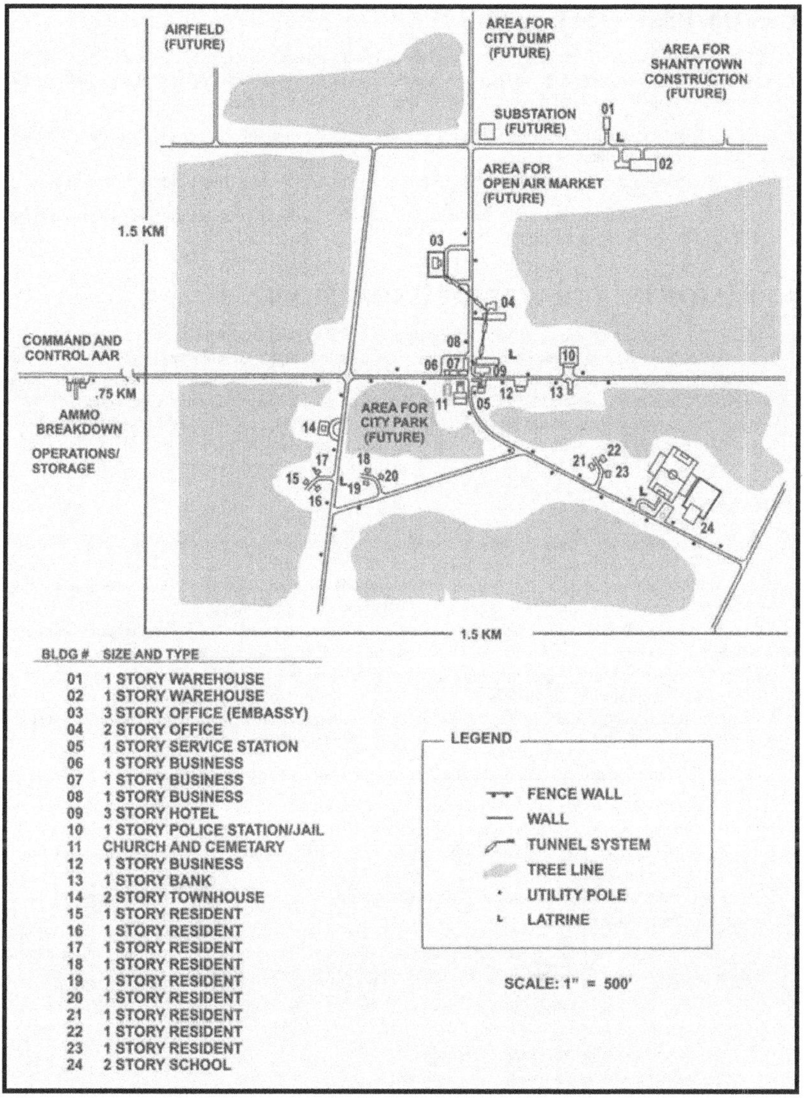

Figure 1-2. Combined Arms Collective Training Facility.

MISSION-ESSENTIAL TASK LIST

1-13. FMs 7-0 and 7-1 explain how to determine a unit's mission-essential task list (METL) and battle tasks, and how to plan, resource, and execute training events. Because training facilities are limited, leaders must use all available time and resources for training and maintaining combat readiness.

- System-specific FMs and TCs teach doctrine, conduct of operations, and employment of weapon systems.
- Tactics, techniques, and procedures (TTP) are lessons learned from training and combat.

1-14. The standards for evaluating training are accessible at the Digital Training Management System (DTMS), https://dtms.army.mil/DTMS.

BATTALION INTEGRATED TRAINING PLAN

1-15. The commander uses the battalion integrated training plan to set goals for training density. He focuses on combined arms operations, emphasizing dismounted Infantry maneuver. Tanks, BFVs, Strykers, attack aviation, and artillery normally provide supporting fires for the dismounted elements. UO training in the CACTF, Shoot House, and Breach Facility should be conducted semiannually and quarterly for the UAC. Army training calendars are divided into two cycles of training: *mission* and *support*. Figure 1-3 and Figure 1-4 show how the urban training strategy and facility usage might be programmed into unit cycles.

RISK ASSESSMENT

1-16. Risk assessment means making operations safe without compromising the level of realistic training (FM 7-1). Commanders continuously assess the risk of training conditions to prevent unnecessary loss of Soldiers and equipment. The degree of risk depends on conditions. Just as this principle applies to other training and warfighting, it also applies to UO. In a training situation, for example, commanders determine whether Soldiers and leaders have conducted this type of training before, whether conducting initial training at night or during limited visibility, and whether Soldiers are tired or exhausted. Risk assessment ensures UO training is not only tough and realistic but also safe. Training enhancers, such as live fire, blank ammunition, simulators, fires, smoke, and other pyrotechnics, must be fully integrated into the training to achieve the training objectives. Soldiers are required to wear earplugs, protective vests, and protective eyewear in order to protect themselves from noise and debris.

1-17. The commander considers risk and safety in urban operations training situations such as--.

- Soldiers engaging in close combat urban operations need to be safety conscious when using blanks, booby traps, and practice grenades. They may be unaware of the exact location of the OPFOR and may be susceptible to OPFOR flash suppressors, booby traps, and practice hand grenades.
- Soldiers should use ladders to enter buildings at upper stories of buildings to prevent unnecessary risks.

1-18. The commander is the safety officer and is responsible for ensuring that his unit, both leaders and Soldiers, comply with safety regulations and the units tactical SOP. All leaders must--.

- Use mission, enemy, terrain, troops, time available, and civilians (METT-TC) factors to identify risks.
- Assess possible losses and their costs.
- Select and develop risk-reduction measures.
- Implement controls by integrating them into plans and orders, SOPs, training performance standards, and rehearsals.
- Supervise and enforce risk reduction measures and safety standards at all times.

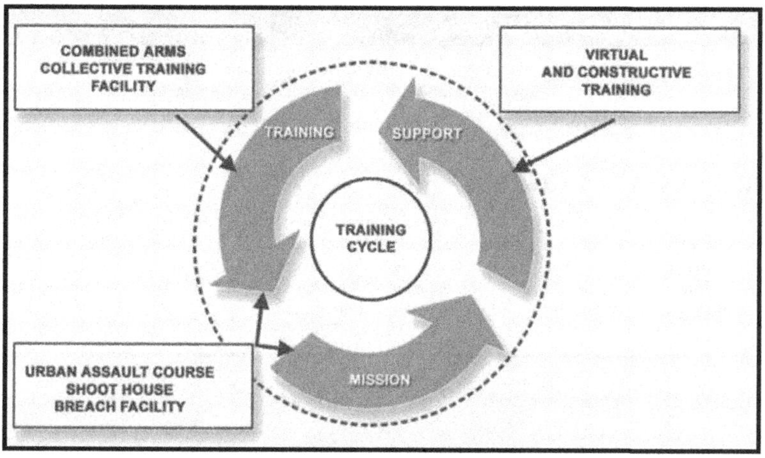

Figure 1-3. Active Army, Home Station, Urban Operations Training Strategy.

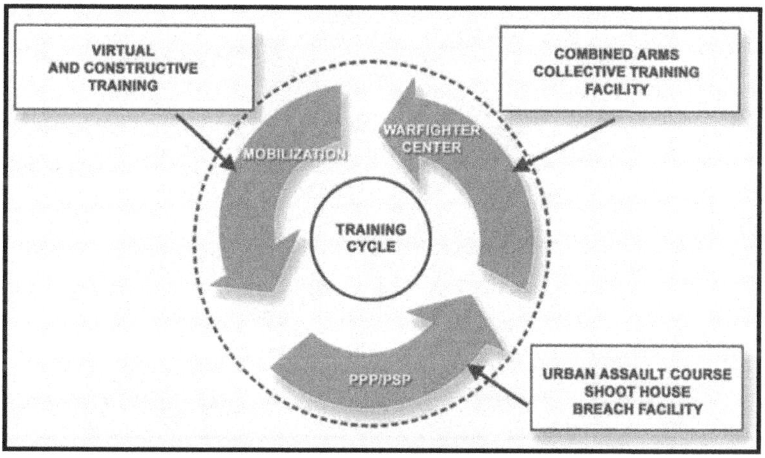

Figure 1-4. Army National Guard Urban Operations Training Strategy.

Section III. TRAINING FACILITIES

Training facilities include three that are discussed in this section: the Urban Assault Course (UAC, Chapter 2), Shoot House (SH, Chapter 3), Breach Facility (BF, Chapter 4). They also include the Combined Arms Collective Training Facility (CACTF, discussed in Chapter 5), Combat Training Centers (CTC), and any other live training that supports the UO training.

URBAN ASSAULT COURSE

1-19. The purpose of the UAC facility is to train a squad-sized force to *Enter a Building and Clear a Room,* improve grenadier gunnery, and validate subterranean TTP. However, the unit's training requirements determine the size of the unit as well as the tasks to be trained. The unit develops its organizational structure and steps for conducting the exercise based on its METL and on their training needs (Figure 1-5). (See Chapter 2 for a more detailed discussion.)

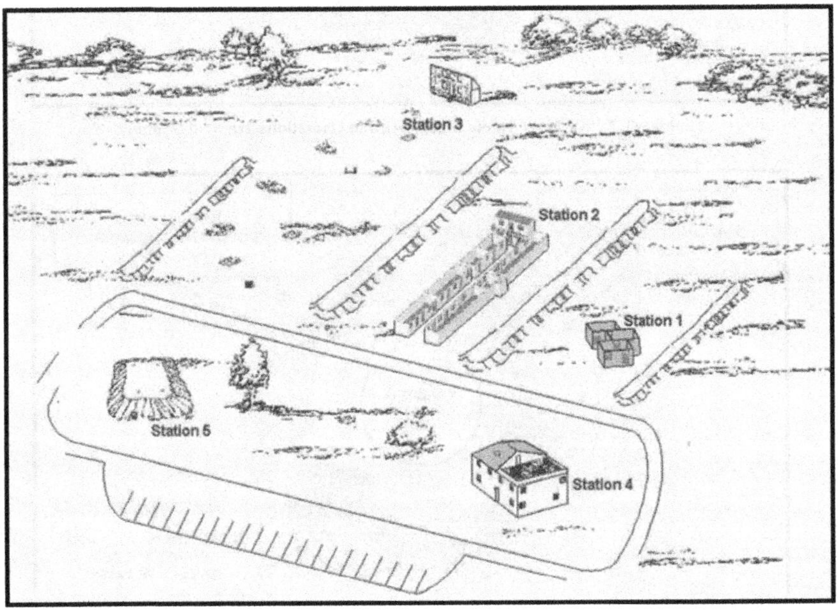

Figure 1-5. Urban Assault Course.

SHOOT HOUSE

1-20. The purpose of this facility is to provide the leader with a facility to train and evaluate fire teams, squads, and platoons during a live-fire exercise (Figure 1-6). Units are trained and evaluated on their ability to conduct tactical movement and engage targets, while practicing target discrimination in an urban environment. (See Chapter 3 for a more detailed discussion.)

Note: The specific designs of the Shoot House and the other facilities may differ from one installation to another, due to variations in training requirements.

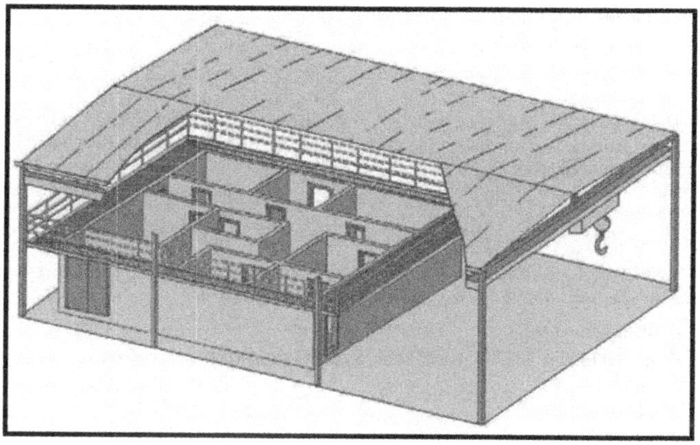

Figure 1-6. Shoot House.

BREACH FACILITY

1-21. The purpose of this facility is to train Warfighters on the technical aspects of mechanical, ballistic, thermal, and explosive breaching techniques (Figure 1-7), including for use in UO. (See Chapter 4 for a more detailed discussion.)

Figure 1-7. Breach Facility.

Section IV. CONDUCT OF TRAINING

Depending on unit proficiency, every training event should follow a *crawl-walk-run* sequence, based on clearly defined task, conditions, and standards. After each training event, leaders conduct an after-action review (AAR).

CRAWL

1-22. The trainer explains each training objective and performance standards, and then talks the Soldiers through the exercise, step by step, describing what each individual or unit must do. This training often includes a demonstration.

WALK

1-23. The Soldiers slowly practice each task to standard. Trainers coach Soldiers at this stage stopping as often as needed to correct mistakes and provide feedback. Soldiers practice each task repeatedly until they can perform it to standard.

RUN

1-24. Soldiers perform each task at full speed as if they are in combat. The Soldiers need feedback during this stage as well. The following can be used to enhance realism:

- Blank ammunition.
- Special Effects Small-Arms Marking System (SESAMS).
- Targetry.
- Pyrotechnics.
- Tactical engagement systems (MILES).
- Opposing forces (force on force).
- Nuclear, biological, and chemical simulation/weapons of mass destruction.
- Training in limited visibility conditions.

AFTER-ACTION REVIEW

1-25. Leaders conduct an after-action review (AAR) after each training event to help Soldiers and units improve their skills by providing immediate feedback. An AAR increases the benefits gained from each training exercise by allowing the leaders and the Soldiers to work together to analyze the performance of each task. Being involved in this analysis enhances the leaders and the Soldiers learning and retention ability. The AAR improves the performance of the leaders and Soldiers in later iterations by allowing interaction and feedback from these individuals during the debrief.

Section V. PREREQUISITE TRAINING

Prerequisite training teaches individuals, crews, units, leaders, and command and staff what they need to know before they are required to perform a task. Prerequisite training builds teamwork and individual confidence at all levels.

COMMAND AND STAFF TRAINING

1-26. The battalion battle staff and leadership must train before the battalion is required to conduct collective training. The battalion command and staff culminates their prerequisite training with a tactical exercise without troops (TEWTs) and command post exercises (CPXs). This sequence of training assists the commander and his staff develop better UO TTP and unit SOP.

LEADER TRAINING

1-27. A comprehensive leader's training program is the key to successful UO training (Appendix D). Time spent teaching and training the battalion leaders (trainers) during the planning and preparation phase will ensure success during the intermediate and culminating phases of training. Realistic training at the unit level will improve the leadership training officers and NCOs receive at the Basic Officer Leadership Course (BOLC), Captains Career Officer Course (CCOC), Basic Noncommissioned Officer Course [BNCOC], and Advanced Noncommissioned Officer Course [ANCOC]. Recommended training should include remedial training on those tasks that the unit had difficulty in during the previous training and evaluation.

INDIVIDUAL TRAINING

1-28. Many individual specific tasks and some collective tasks can be taught and practiced in garrison while most collective tasks are best conducted at the installation UO facilities and weapons ranges. The unit's preliminary individual training tasks should focus on three general areas: urban combat skills, urban marksmanship skills, and physical fitness. The following are example preliminary training tasks or TTP:

- Individual movement techniques.
- Battle drills.
- Weapon positioning in an urban area.
- Roadblock and vehicle search procedures.
- Dismounted and mounted urban navigation.
- Urban scanning, quick-fire, and assault fire techniques.
- Hasty urban firing positions.
- Preparation of urban firing positions.

1-29. Prior to urban collective training cycle, trainers should develop urban-specific firing conditions in conjunction with weapons firing ranges. Build facades on existing ranges, or use the unit's local UAC (if available). Urban combat skills include--

- Firing positions behind rubble, around corners, from windows or rooftops.
- Urban targets behind rubble, in windows and doorways.
- Fire commands.
- Lifting and shifting fires from lower to upper story windows.
- Fire control and fire discipline.
- Rapid firing engagements.
- ROE firing scenarios.

1-30. Urban physical fitness skills include--

- Endurance runs and upper body conditioning.
- Vaulting exercises.

- Urban-specific obstacle course negotiation.
- Leadership Reaction Course negotiation.

INDIVIDUAL TASK TRAINING

1-31. Urban operations require the Soldier to be proficient in those individual tasks that are unique to urban combat. Units should incorporate into their training regime those urban fighting TTP that have proven successful based on lessons learned from combat operations.

Task Force Soldier--Warrior Common Tasks, Skill Levels 1-3:

- 071-326-0541, *Perform Movement Techniques during an Urban Operation.*
- 071-326-0542, *Enter a Building during an Urban Operation.*
- 071-326-0550, *Prepare Positions for Individual and Crew-Served Weapons during an Urban Operation.*
- 071-326-0557, *Select Hasty Firing Positions during an Urban Operation.*
- 071-440-0027, *Execute Mechanical Breach in an Urban Operation.*
- 071-440-0028, *Engage Targets during an Urban Operation.*
- 071-440-0029, *Conduct a Breach during an Urban Operation.*
- 071-440-0003, *Conduct an Attack on a Building by a Squad during an Urban Operation.*
- 071-440-0006, *Conduct a Defense by Squad during an Urban Operation.*

FM 3-06.11, *Combined Arms Operations in Urban Terrain*

Movement Techniques

- Crossing of a wall.
- Movement around corners.
- Movement past windows.
- Use of doorways.
- Movement parallel to buildings.
- Crossing open areas.
- Fire team employment.
- Movement between positions.
- Movement inside a building.

Entry Techniques

- Upper building levels.
- Use of ladders.
- Scaling of walls.
- Entry at lower levels.
- Hand grenades.

Firing Positions

- Hasty firing position.
- Prepared firing position.
- Target acquisition.
- Employment of snipers.

COLLECTIVE TASK TRAINING

1-32. Specific FMs teach doctrine, how to conduct operations, and how to employ weapon systems. TTP are lessons learned from the training and combat. The standards for conducting and evaluating collective training are accessible at the Digital Training Management System (DTMS), https://dtms.army.mil/DTMS. Those tasks that do not specifically address UO must be modified for use in an urban environment.

RESOURCES

1-33. Field manuals for the squad and platoon are--:

- FM 3-21.8, *Infantry Rifle Platoon and Squad.*
- FM 3-21.71, *Mechanized Infantry Platoon and Squad (BFV).*
- FM 3-21.9, *SBCT Infantry Rifle Platoon and Squad.*

1-34. Field manuals for the company are--.

- FM 3-21.10, *IBCT Rifle Company.*
- FM 3-21.11, *SBCT Rifle Company.*

1-35. Field manuals and mission training plans for the battalion are--.

- FM 3-21.20, *IBCT Infantry Battalion.*
- FM 3-21.21, *SBCT Infantry Battalion.*
- FM 3-06.11, *Combined Arms Operations in Urban Terrain.*
- TC 7-98-1, *Brigade and Battalion Stability and Support Operations.*

Chapter 2

Urban Assault Course

This chapter describes the Urban Assault Course (UAC), which incorporates doctrine and tactics from FM 3-06.11. The UAC contains five stations:

- Individual and Team Task/Technique Trainer Station 1
- Squad and Platoon Task/Technique Trainer Station 2
- Grenadier Gunnery Trainer Station 3
- Urban Offense/Defense Building Station 4
- Underground Trainer Station 5

These stations are designed for individual tasks and small-unit collective task training. Various types of organizations with a variety of missions can train at the UAC. Tactical operations in an urban environment may include combat to stability and support tasks. (Appendix C provides a SASO unit training matrix). The UAC is a dismounted training facility.

Note: The station discussions are arranged so that they may be removed from the book for ease of use.

PURPOSE

2-1. The purpose is to provide squad and platoon size units with a facility to train and evaluate urban operations tasks (Figure 2-1). Only Station 3, Grenadier Gunnery, is intended for live-fire training. It supports use of 40-mm TP and 5.56-mm service ammunition.

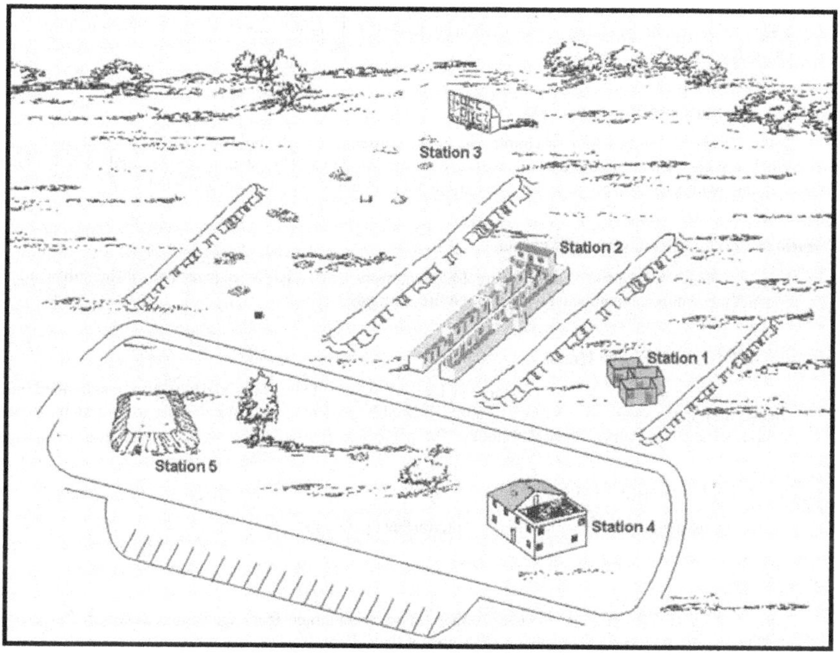

Figure 2-1. Urban Assault Course.

DESCRIPTION

2-2. The UAC has five training stations that facilitate the ***crawl-walk-run*** training concept:

Station 1, Individual and Team Trainer

2-3. This station has three rooms where team leaders and squad leaders train the basics of building and room clearing.

Station 2, Squad and Platoon Trainer

2-4. This station has four structures and multiple rooms. Squads build on tasks learned at Station 1, and begin to learn the concepts of clearing multiple buildings. This station can be used as individual buildings with a narrow street, or as rooms inside a building with a long, connecting hallway.

Station 3, Grenadier Gunnery Trainer

2-5. This is a live-fire station where M203 gunners master target engagements in an urban area, conduct tactical movement, and respond to the fire commands.

Station 4, Offense/Defense House

2-6. This is where a platoon trains to attack or defend a building. This station can also be divided into smaller training stations to reinforce training. It is also used to train tasks not yet trained at the other stations such as upper-level entry techniques.

Station 5, Underground Trainer

2-7. This station provides training for subterranean operations.

TARGETRY

2-8. Targets throughout the UAC are either 3D precision targets (for engagements less than 50 meters) or 2D nonprecision targets (for engagements greater than 50 meters).

Precision Targets

2-9. Precision targets support short distance engagement sensors. These sensors, located inside the precision targets, respond only to lethal shot placements.

- Torso lethal zone .... 18" wide x 8" high
- Neck-head lethal zone .... 4" wide x 8" high

Nonprecision Targets

2-10. Nonprecision targets support engagements at ranges greater than 50 meters. These targets are 2D pop-up type targets.

Training Audience

2-11. The facility gives Infantry, Engineer, Military Police, Cavalry, and other units a facility for training individual Soldiers through platoon-level UO skills.

Training Information

2-12. Combat units can use this facility as a mission rehearsal exercise site. Recommended training frequency is quarterly. For each UAC station, they should provide the purpose, a description, targetry, controller and safety requirements, individual UO tasks to be trained, training support requirements, and conduct of training information, if applicable:

Purpose

2-13. This UAC station [*state purpose of station*]. [*Units may use the station in the manner originally intended, or they may modify it to meet their unique training needs.*]

Description

2-14. This station is [*describe station use and design*].

Targetry

2-15. Recommended targetry for training at this station includes [*describe*]. Targets may be moved, modified, or added, depending on available target systems or training requirements.

Controller

2-16. Controller requirements for this station include [*describe*].

Safety

2-17. Safety requirements and considerations for this station include [*describe*]. [*The unit or installation range SOP may add more stringent safety requirements.*]

Individual Tasks or Skills Trained

2-18. These may be found in the Digital Training Management System located online at https://dtms.army.mil/DTMS.

Training Support

2-19. Minimum materiel requirements to support training at this station include [describe].

Conduct of Training

2-20. Example training scenarios and sequence of events for this station include [describe]. [The UAC can accommodate an assortment of training ammunition and pyrotechnics.]

STATION 1: INDIVIDUAL AND TEAM TASK/TECHNIQUE TRAINER

2-21. This station is used to train individual and collective tasks, tactics, techniques, and procedures, specifically *Enter a Building and Clear a Room* and *Engage Targets* (Figure 2-2). It has three rooms where team leaders and squad leaders train the basics of building and room clearing.

DESCRIPTION

2-22. Station 1 has a single wood structure. Its three rooms have one or more open doorways and windows, and a man-sized hole.

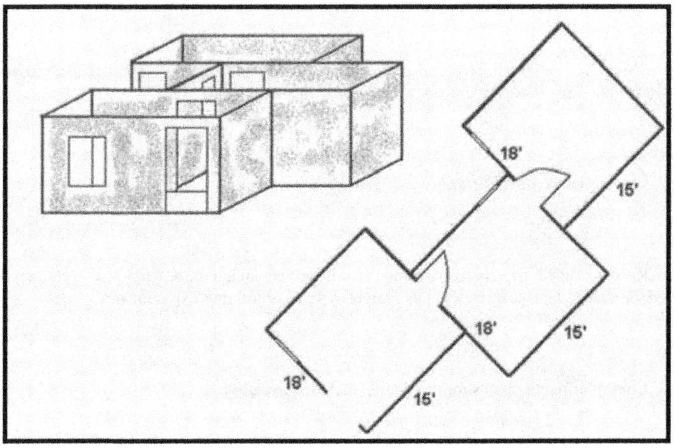

Figure 2-2. Individual and team task/technique trainer, Station 1.

TARGETRY

2-23. Targets present realistic close quarters combat situations (0 to 15 meters), which consist of a possible mixture of combatant and noncombatant targets. Target discrimination is trained by presenting up to six precision targets. Targetry sensors respond only to lethal shot placement. They also aid in training during limited visibility.

CONTROLLER REQUIREMENTS

2-24. These are based on unit requirements.

SAFETY REQUIREMENTS

2-25. Only smoke and practice grenades, or M84 stun grenades, may be used at this station.

> **WARNING**
>
> **To avoid injury, place the weapon on SAFE when climbing through obstacles such as windows, fences, and walls during training.**

INDIVIDUAL/COLLECTIVE TASKS TRAINED

2-26. This is noninclusive.

Collective Task

2-27. *Enter a Building/Clear a Room.*

Supporting Individual Tasks

- 071-311-2007, *Engage Targets with M16-Series Rifle/M4 Carbine Rifle.*
- 071-315-2308, *Engage Targets with M16-Series Rifle/M4 Carbine Using a Night Vision Sight AN/PVS-4.*
- 071-010-0006, *Engage Targets with M249 Machine Gun.*
- 071-325-4407, *Employ Hand Grenades.*
- 071-326-0503, *Move Over, Through, or Around Obstacles (except Minefields).*
- 052-193-1013, *Neutralize Booby Traps.*
- 071-326-0501, *Move as a Member of a Fire Team.*
- 071-326-0541, *Perform Movement Techniques during an Urban Operation.*
- 071-326-0557, *Select Hasty Firing Positions during an Urban Operation.*
- 071-326-5605, *Control Movement of a Fire Team.*
- 071-326-5611, *Conduct the Maneuver of a Squad.*
- 071-710-0008, *Operate Night Vision Goggles AN/PVS-7B.*

Related Tactics, Techniques, and Procedures

- FM 3-06.11, *Combined Arms Operations in Urban Terrain.*
- FM 3-22.9, *Rifle Marksmanship.*

TRAINING SUPPORT REQUIREMENTS

2-28. Use Table 2-1 for the minimum ammunition required to train at this station.

Table 2-1. Support requirements for Station 1.

Ammunition	*Rounds per Weapon*
5.56-mm Blank	10 rounds
5.56 mm Blank (Linked)	15 rounds
Practice Grenade Bodies	2 per team
Practice Grenade Fuses	2 per team

CONDUCT OF TRAINING

2-29. Table 2-2 lists the training sequence and times for Station 1.

Table 2-2. Sequence of events and times for Station 1.

Event	*Action*	*Estimated Time*
1	*Employ Hand Grenades*	30 minutes crawl, walk, run
	Perform Movement Techniques • *Engage Target with M16* • *Target Discrimination*	20 minutes crawl 20 minutes walk 10 minutes run
2	*Enter a Building*	20 minutes crawl 20 minutes walk 10 minutes run
3	*Enter and Clear a Room*	20 minutes crawl 20 minutes walk 10 minutes run
4	*Conduct an AAR*	40 minutes

Example Scenarios, Assault Team, Varied Entry Positions

2-30. Figures 2-3 and 2-4 show example scenarios. Rather than using the examples, units should tailor their training events based on their unit's METL and training needs.

Varied Entry Positions

2-31. The team leader first trains to execute *Varied Entry Positions* without targets, which is part of the *crawl* phase of training. The trainer has the assault teams position at various places to demonstrate how the start position can affect the final points of domination. (Figure 2-3).

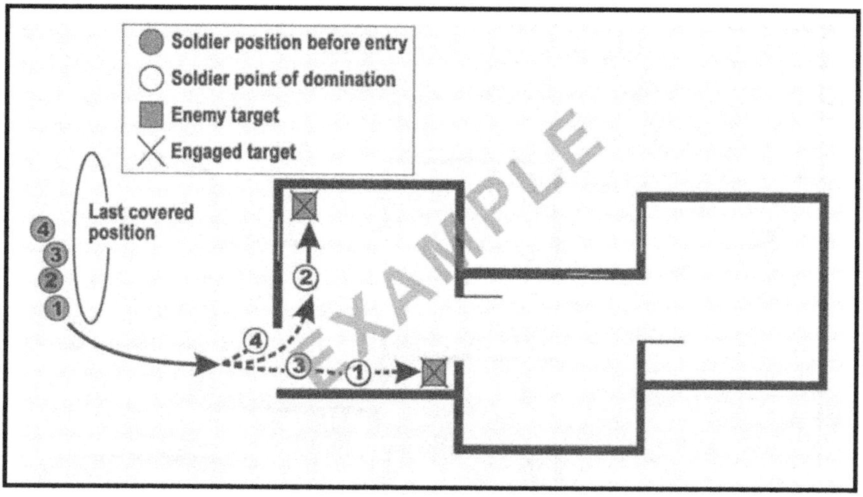

Figure 2-3. Example Station 1 scenario, *Varied Entry Positions*.

Multiple Room-Clearing Techniques

2-32. The team leader trains his Soldiers to execute *Varied Entry Positions* using targets in a multiple room scenario. This is part of the *walk* phase. The trainer has the assault teams enter the room and engage targets. He has them use using precision marksmanship techniques with the Tactical Engagement System (TES) or SESAMS (Figure 2-4). The assault team enters the first room and engages the targets. As the number two Soldier clears his area of responsibility, he sees and engages the target in the second room, but remains in place until the assault teams finishes clearing the first room. Once the room is clear, the assault team moves to the second room and continues clearing, or lets the next team move through.

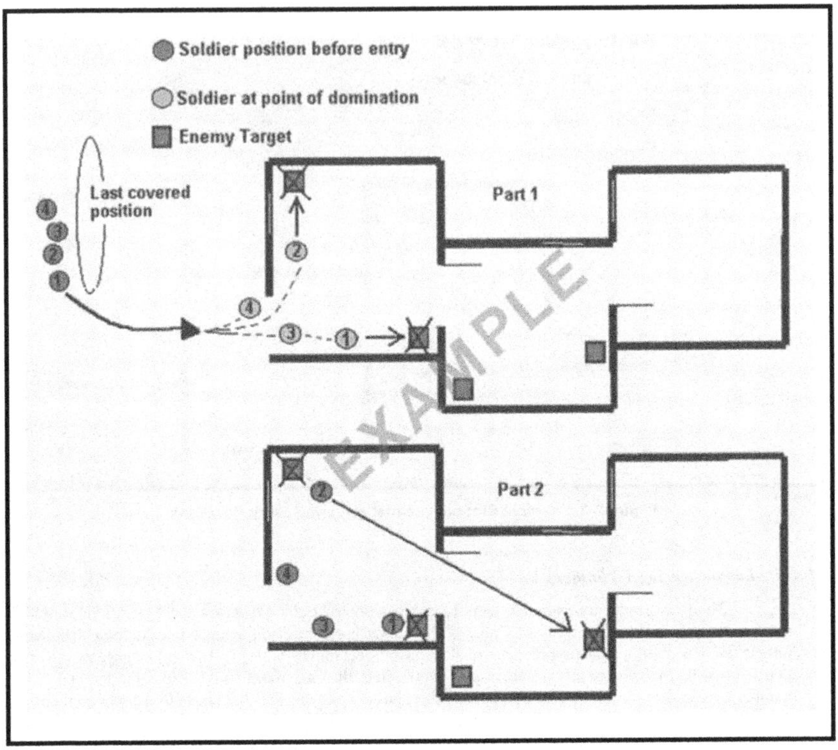

Figure 2-4. Example Station 1 scenario,
Multiple Room-Clearing Techniques.

STATION 2: SQUAD AND PLATOON TASK/TECHNIQUE TRAINER

2-33. This station has four structures and multiple rooms. Squads build on tasks learned at Station 1, and begin to learn the concepts of clearing multiple buildings. This station can be used as individual buildings with a narrow street, or as rooms inside a building with a long, connecting hallway.

2-34. Station 2 is used to train individual and collective tasks, tactics, techniques, and procedures, specifically, *Move Tactically Down a Street (Hallway)*, *Enter a Building and Clear a Room*, and *Engage Targets*. Station 2 increases the complexity of command, control, and maneuver (Figure 2-5).

DESCRIPTION

2-35. This station has four separate structures that face inward to create a street. It also introduces a two-story building.

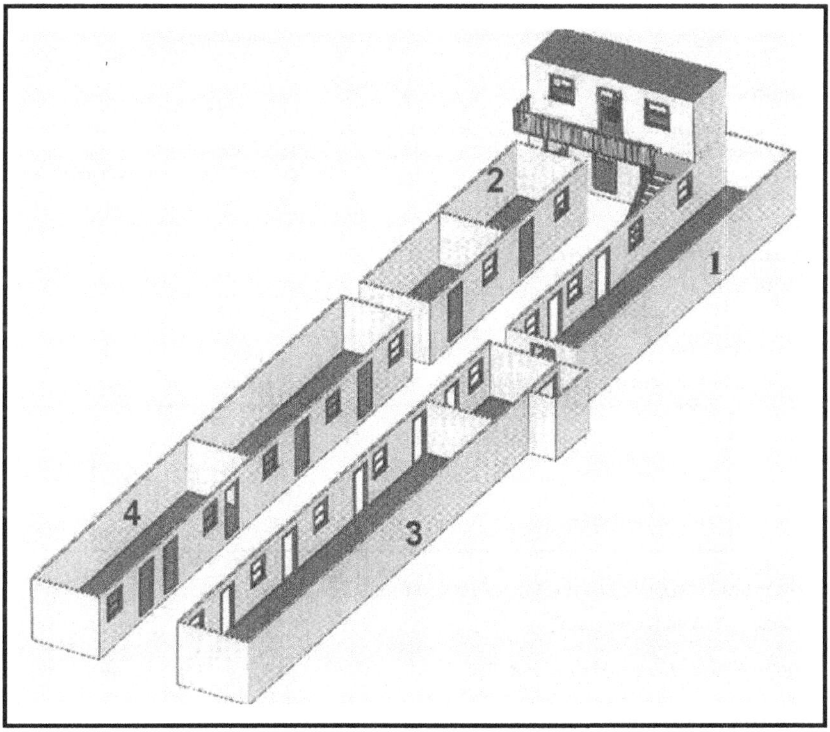

Figure 2-5. Squad and platoon task and technique trainer, Station 2.

> Building 1 is L-shaped.
>
> - The long leg of the building to the right looks like a single-story building.
> - The short leg of the building to the front has two floors. A landing along the length of the second floor has outside stairs. An inside stairway also joins the floors.
> - The two legs of the "L" are adjacent, but without interior access to each other.
>
> Building 2 has a single floor and two rooms, each with a door and window. An open area much like a street separates this structure from the first.
>
> Building 3 is a row of wooden buildings along a street. It has many windows and doorways. At the far end is an adjoining room.
>
> Building 4 is on the left. It mirrors Building 3, plus it has a wall on the side at the near end.

2-36. Units can place additional walls in any structure by using target cloth or other material. They can do this to add to the number of rooms or to increase the complexity of the training (Appendix A).

Targetry

2-37. Station 2 targets present realistic close quarters combat situations (0 to 15 meters), with the option to customize the mixture of combatant and noncombatant targets. Target discrimination is trained on up to ten precision targets. Targetry sensors respond only to lethal shot placement, and facilitate training during limited visibility.

Controller Requirements

2-38. These are based on unit requirements.

Safety Requirements

2-39. At this station, Soldiers may only use smoke grenades, practice grenades, and M84 stun grenades.

> **WARNING**
>
> **To avoid injury, place the weapon on SAFE when climbing through obstacles such as windows, fences, and walls during training.**

Individual and Collective Tasks Trained

2-40. This list is noninclusive.

Collective Task--*Enter and Clear a Building*

Related Collective Tasks

- *Conduct Consolidation and Reorganization (Company/Platoon).*
- *Treat Casualties.*
- *Evacuate Casualties.*
- *Defend MOUT Building.*

- *Conduct Tactical Movement.*
- *Enter and Clear a Room.*

Supporting Individual Tasks

- 052-193-1013, *Neutralize Booby Traps.*
- 071-010-0006, *Engage Targets with M249 Machine Gun.*
- 071-311-2007, *Engage Targets with M16-Series Rifle/M4 Carbine Rifle.*
- 071-315-2308, *Engage Targets with M16-Series Rifle/M4 Carbine Using a Night Vision Sight AN/PVS-4.*
- 071-325-4407, *Employ Hand Grenades.*
- 071-326-0501, *Move as a Member of a Fire Team.*
- 071-326-0503, *Move Over, Through, or Around Obstacles (except Minefields).*
- 071-326-0541, *Perform Movement Techniques during an Urban Operation.*
- 071-326-0557, *Select Hasty Firing Positions during an Urban Operation.*
- 071-326-0608, *Use Visual Signaling Techniques.*
- 071-326-5605, *Control Movement of a Fire Team.*
- 071-326-5611, *Conduct the Maneuver of a Squad.*
- 071-710-0008, *Operate Night Vision Goggles AN/PVS-7B.*
- 113-571-1022, *Perform Voice Communications.*
- 191-377-5250, *Handle Enemy Personnel and Equipment.*

Related Tactics, Techniques, and Procedures

- FM 3-06.11, *Combined Arms Operations in Urban Terrain.*
  -- Chapter 3, *Offensive Operation.*
  -- Chapter 5, *Fundamental Combat Skills.*
  -- Appendix M, *Field-Expedient Breaching of Common Urban Barriers.*
- FM 3-22.9, *Rifle Marksmanship.*

TRAINING SUPPORT REQUIREMENTS

2-41. Table 2-3 shows the minimum ammunition required to train at this station.

Table 2-3. Support requirements for Station 2.

Ammunition	*Rounds per Weapon*
5.56-mm blank	16
5.56-mm blank (linked)	24
Practice grenade fuses	8 per squad
Practice grenade bodies	8

CONDUCT OF TRAINING

2-42. Table 2-4 shows the sequence of events and times for Station 2. Figure 2-6 through Figure 2-28 show example scenarios. However, rather than use these examples, the unit should customize its training events based on its METL and training needs.

Table 2-4. Sequence of events and times for Station 2.

Event	*Action*	*Estimated Time*
1	*Conduct Tactical Movement in a Built-Up Area* *Move Across an Open Area* *Move Parallel to Buildings*	10 min Crawl 20 min Walk 10 min Run
2	*Enter and Clear a Room (Squad)* *Clear a Stairwell* *Clear a Hallway* *Conduct a Breach*	10 min Crawl 20 min Walk 10 min Run
3	*Enter and Clear a Building (Platoon)* *Clear Multiple Rooms* *Consolidate and Reorganize*	10 min Crawl 20 min Walk 10 min Run
4	*Conduct AAR (Squad)*	10 min
NOTE: Training should include techniques outlined FM 3-06.11.		

2-43. The squad is clearing Station 2 as if it were a single building, and the street a hallway. Fire Team A moves down the left side, throws the appropriate grenade IAW ROE, enters, engages the target, and clears the room on that side. The last Soldier remains outside and secures the rear Fire Team B prepares to enter the next room (Figure 2-6).

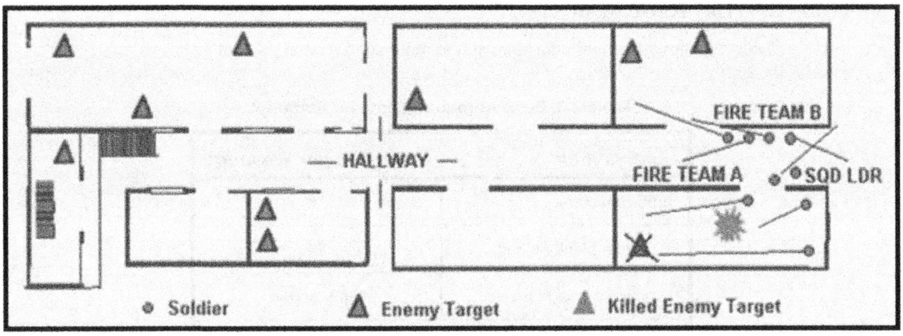

Figure 2-6. Conducting tactical movement.

2-44. Fire Team B throws the appropriate grenade, enters, engages the targets, and clears the room on the right while the last Soldier secures the rear Fire Team A prepares to continue movement to the next room (Figure 2-7).

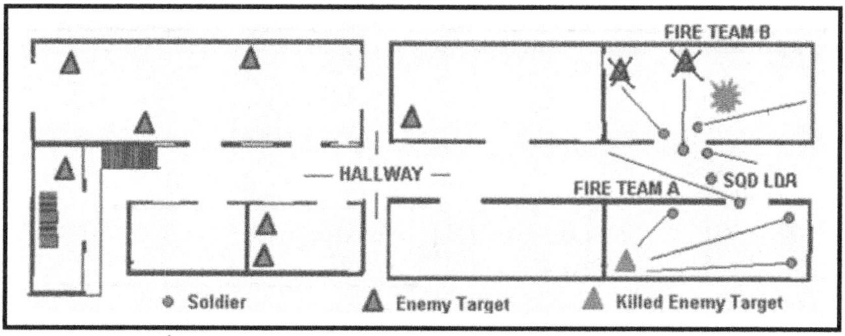

Figure 2-7. Team B--Clearing a building.

2-45. Fire Team A moves across the hall, throws the appropriate grenade, enters, engages the target, and clears the next room Three team members enter the room, the fourth remains in the doorway and secures the hallway (Figure 2-8).

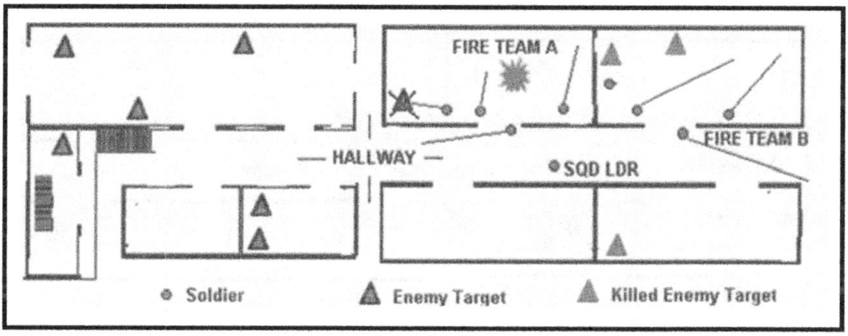

Figure 2-8. Team moving across the hall.

2-46. Fire Team B moves in a cross-cover formation down the hallway to the next room, throws the appropriate grenade, and clears the room. As Fire Team B passes the security man of Team A, he turns and becomes rear security. The last Soldier in Fire Team B remains by the door and secures the hallway (Figure 2-9).

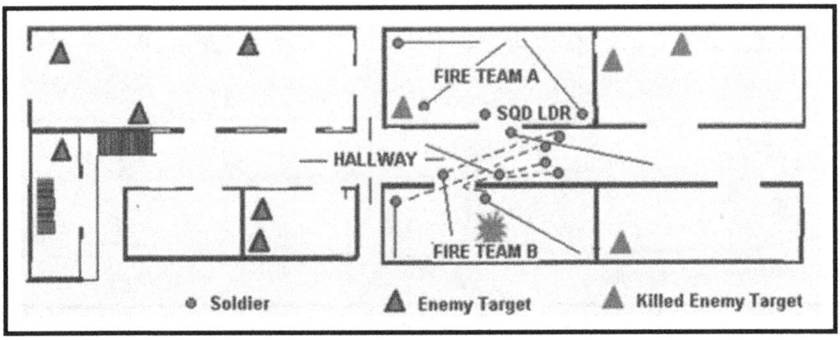

Figure 2-9. Team B--Moving in cross cover formation.

2-47. Fire Team A clears the hallway intersection, while Team B uses the Rolling-T technique to pass through Team A en route to the next room (Figure 2-10).

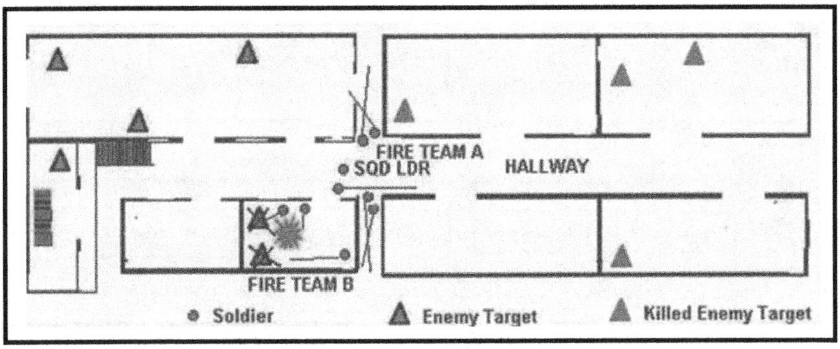

Figure 2-10. Team A--Clearing a hallway intersection.

2-48. Fire Team B throws the appropriate grenade, enters, engages the targets, and clears the room. Only three Soldiers enter, while the fourth Soldier provides rear security. Fire Team A prepares to move (Figure 2-11).

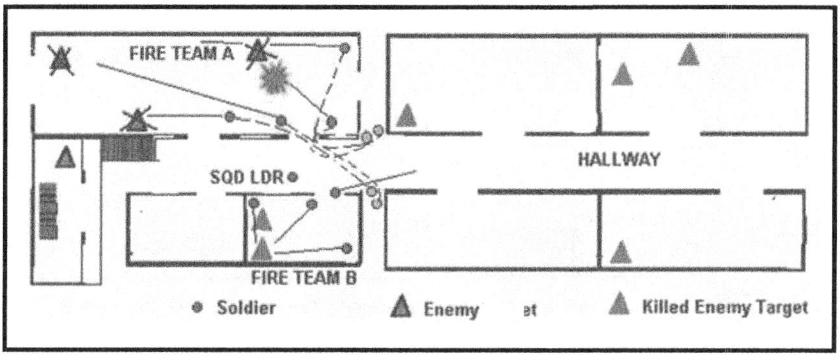

Figure 2-11. Team B--Clearing a room.

2-49. Fire Team A moves from the intersection and throws the appropriate grenade. The team enters, engages the targets, and then clears the next room. All team members enter, because of the size of the room. Fire Team B prepares to move (Figure 2-12).

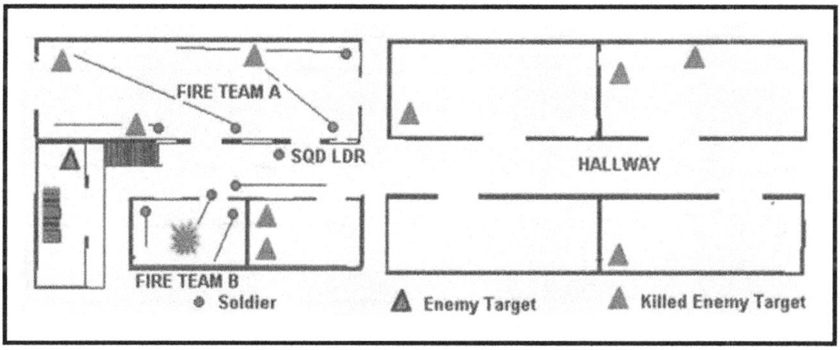

Figure 2-12. Team A--Clearing room by room.

2-50. Fire Team B employs the appropriate grenade, enters, and engages the target. Three Soldiers clear the next room. The fourth remains in the doorway and secures the rear. Fire Team A prepares to move up the stairs (Figure 2-13).

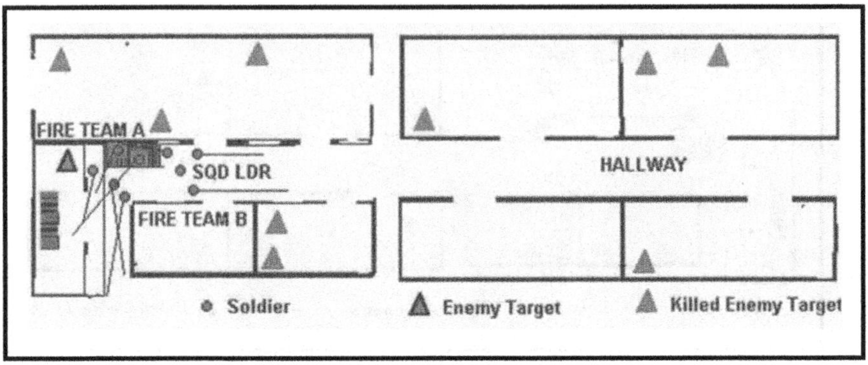

Figure 2-13. Team B--Clearing the next room.

2-51. Fire Team A moves up the stairs and onto the landing. The team prepares to enter the room off the landing. Fire Team B splits, and two members move around the corner and secure the area across from the stairs on the lower level. The other two members of Team B secure the hallway the squad came from (Figure 2-14).

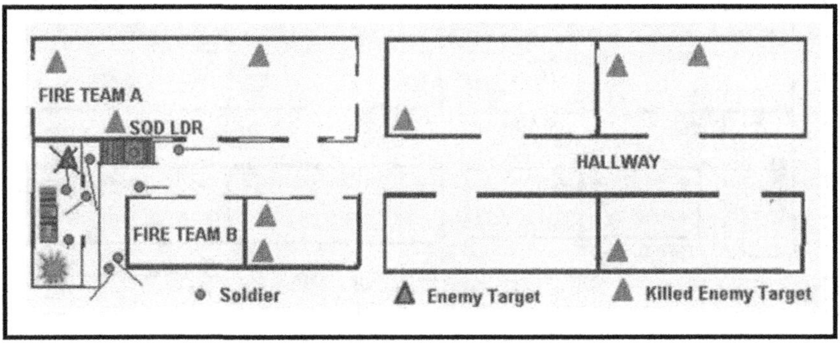

Figure 2-14. Securing the stairs and hallway.

2-52. Fire Team A members employ the appropriate grenade, enters the room, engages the target, and clears it. They find another set of stairs, and prepare to move up them. Fire Team B continues to secure the rear. The other team members move to the end of the short hallway and secure both directions (Figure 2-15).

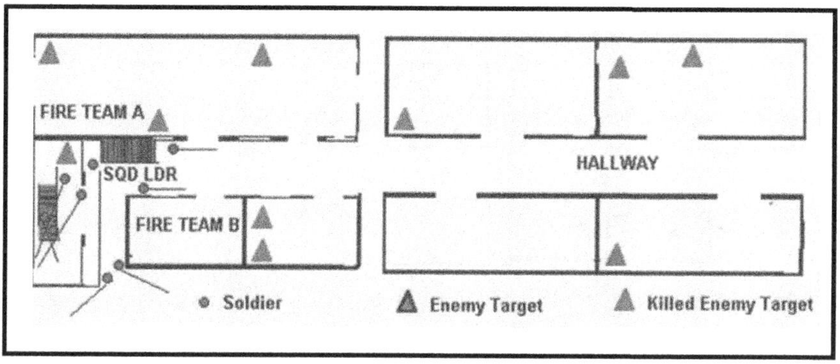

Figure 2-15. Preparing to clear more stairs.

2-53. Fire Team A clears the stairs and the landing. Fire team B continues to secure the ground level, ready to assist Team A, if necessary (Figure 2-16).

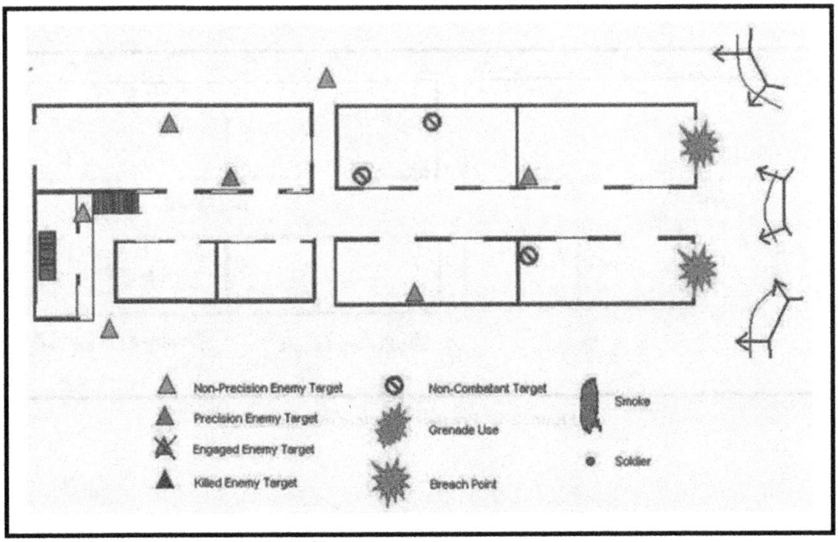

Figure 2-16. Team A--Clearing stairs and landing.

2-54. The previous example used no noncombatant targets, but used blanks/SESAMS and TES. This technique is the *crawl* phase for this station. Once the squad is ready to move on to more challenging exercises, the platoon leader/sergeant can add targets into the scenario. As proficiency increases, mixed combatant and noncombatant targets can be added. At any point, the platoon leader/sergeant can, until the platoon achieves proficiency, include multiple squads on the station. This increases the complexity of the scenarios.

Example Scenario--Preparing to Gain a Foothold

2-55. In this example platoon scenario, noncombatants are on the battlefield. The platoon has decided to clear through the buildings. They are going to try to do this without using the streets to move from building to building. Two squads will assault, and the other will remain in support/reserve. The support element isolates the buildings, while the assault team for each squad prepares and breaches each building in order to gain a foothold (Figure 2-17).

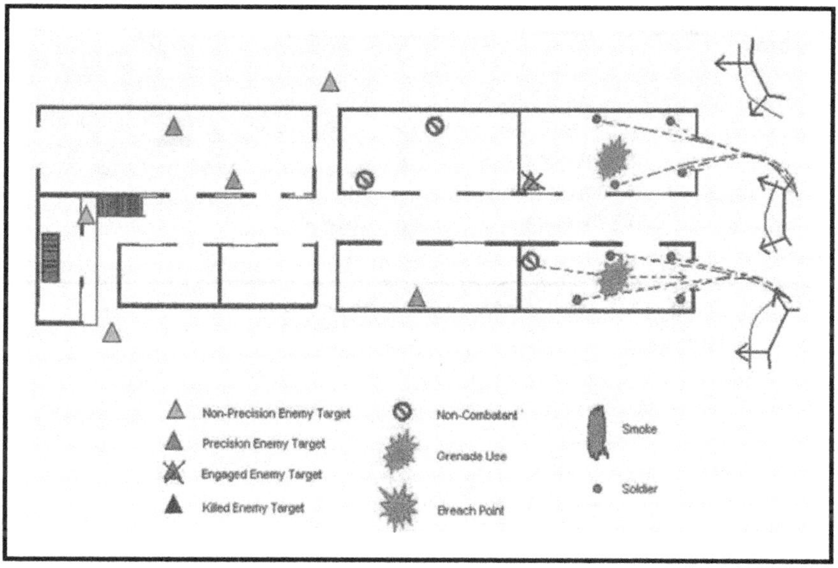

Figure 2-17. Preparing to gain a foothold.

2-56. The assault teams throw a grenade into the breach and their members follow, engaging enemy targets. They secure and move noncombatants to the collection point (CP) (Figure 2-18).

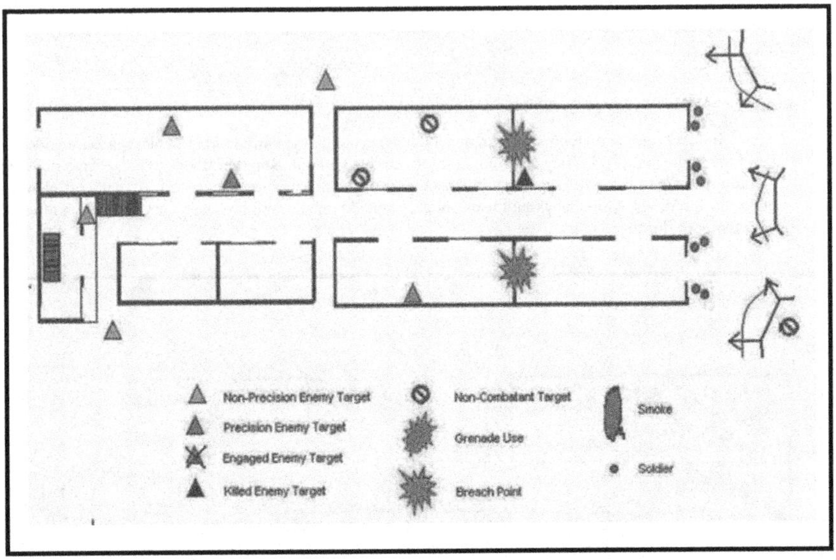

Figure 2-18. Securing and moving noncombatants to the command post.

2-57. The teams prepare and detonate wall breaches to continue movement through the building. Once the charges are ready for detonation, the assault teams move to a covered position and execute the breach (Figure 2-19).

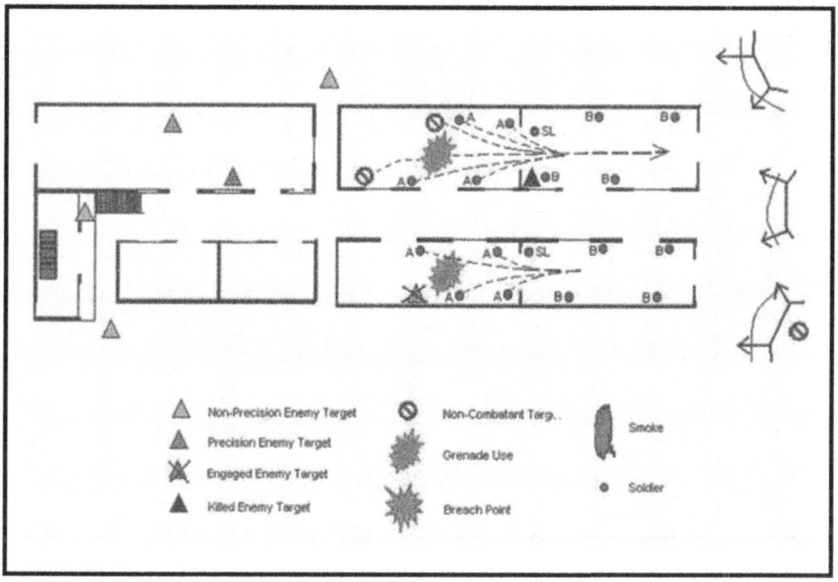

Figure 2-19. Executing breach.

2-58. The assault team employs grenades into the next room, enters the room, eliminates the enemy, and then secures and moves the noncombatants to a designated safe area. At the same time, the next assault team moves into the foothold (Figure 2-20).

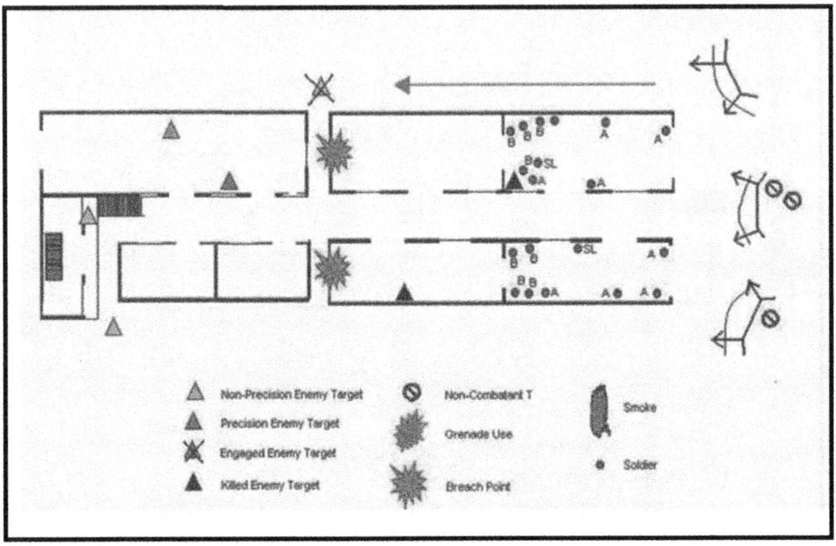

Figure 2-20. Moving to the foothold.

2-59. With the second room secured, another assault team moves forward and prepares to breach the wall Once the charges are placed, the teams move back to a covered position before detonating the charge. With the breach a success, the first nonprecision target is exposed and engaged by the support element (Figure 2-21).

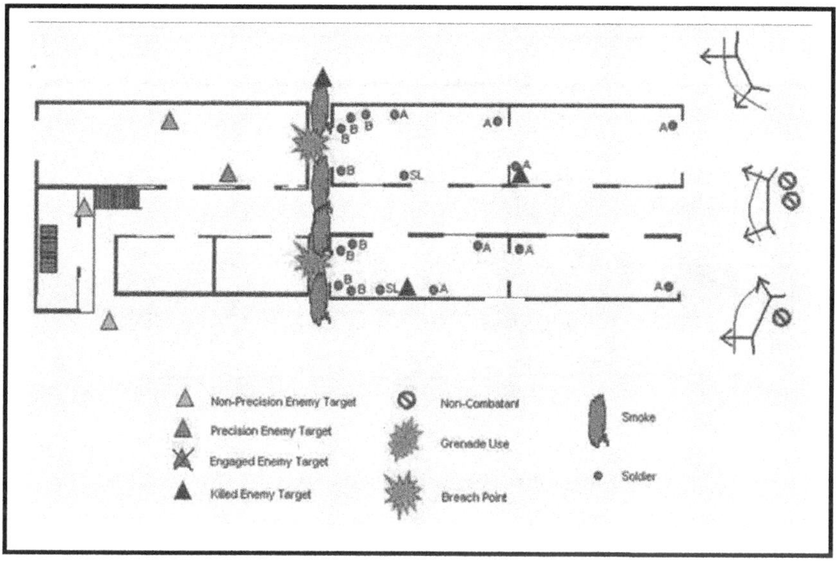

Figure 2-21. Breaching the wall.

2-60. Upon inspecting the breach, the assault teams find that they have an alley with another wall to breach. They employ smoke to obscure the teams placing the breaching charge. The first assault teams take up securing positions throughout the already cleared areas. Follow-on assault teams move to a covered position and execute the breach (Figure 2-22).

Note: Use smoke only in the alley, not inside the buildings.

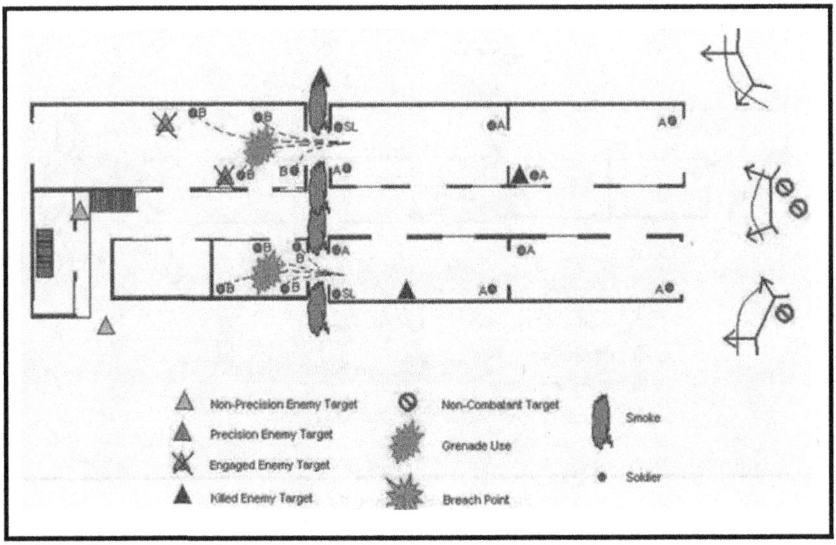

Figure 2-22. Securing a wall in an alley.

2-61. Once the assault teams execute the breach, they must determine if they need to use more smoke to cover their movement. Either way, once they have enough smoke to cover their movement, they throw in a grenade, enter, engage enemy targets, and clear the rooms (Figure 2-23).

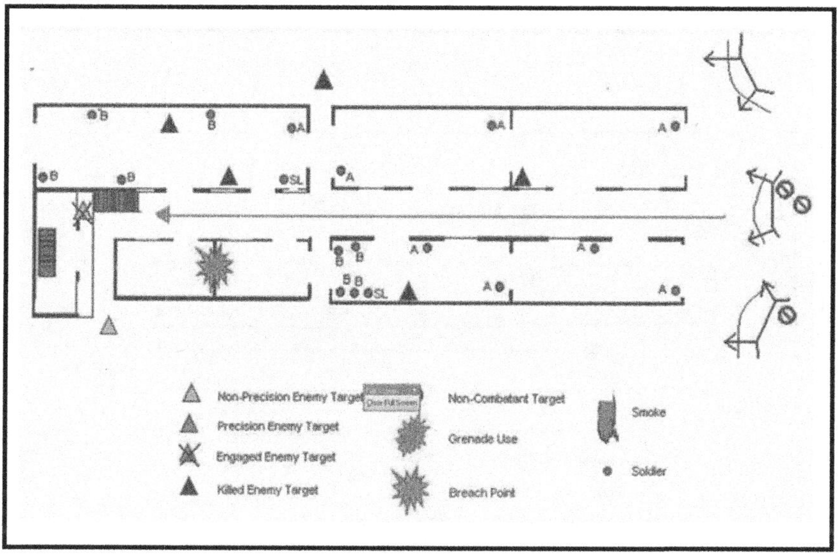

Figure 2-23. Obtaining obscurity.

2-62. While one team finds itself at the end of the building in their sector, the other team is faced with another wall and more clearing. One team prepares to breach the wall, and the other team takes up a hasty position and prepares for a counterattack. When the teams execute the breach, the next nonprecision target presents, and the support element engages it (Figure 2-24).

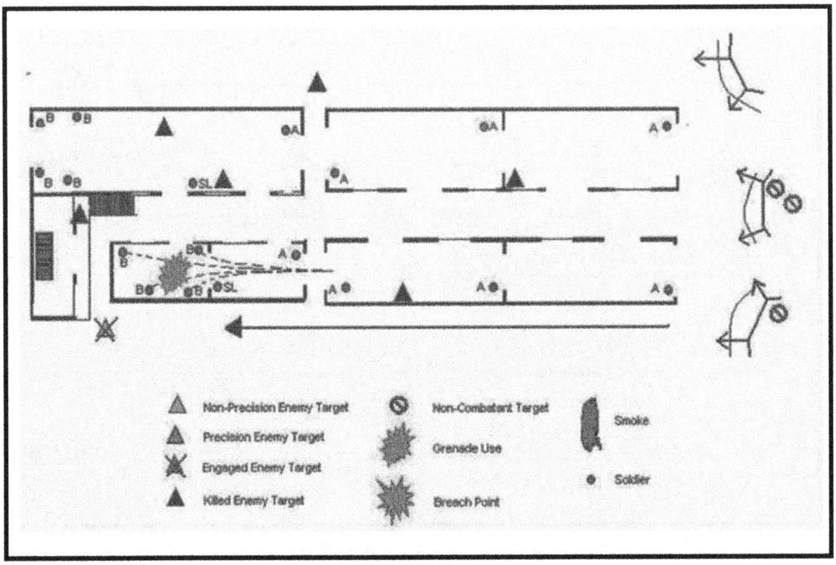

Figure 2-24. Clearing the wall.

2-63. Someone throws a grenade through the breach. The assault team enters and secures the room. When the grenade explodes inside the room, the final nonprecision target presents, and the support element engages it (Figure 2-25).

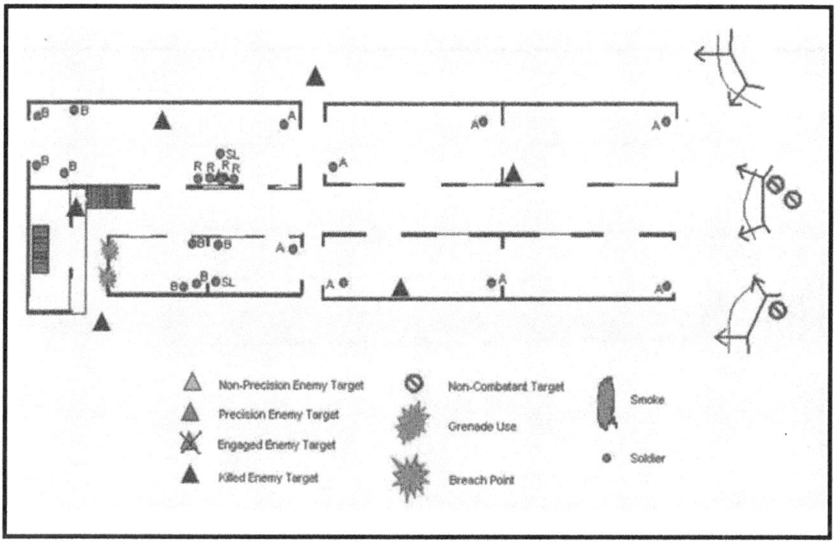

Figure 2-25. Securing a room.

2-64. With the first squad reaching its limit of advance, and with the second squad reaching the end of the building, the platoon leader calls another assault team forward from the reserve. The reserve assault team moves forward through the buildings, and positions itself to clear up the outside stairs of the two-story structure. The second squad places and detonates loophole charges on the far wall of the building. This supports the clearing of the last building (Figure 2-26).

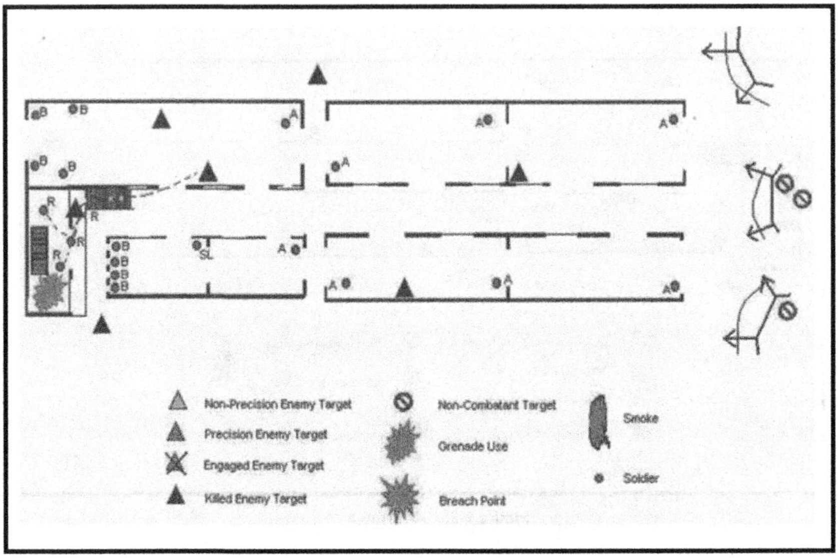

Figure 2-26. Reserve assault team moving forward.

2-65. Team B of the second squad moves into supporting fire positions. The new assault teams move up the outside stairs to the second floor of the final structure. They clear the balcony and the first room on the second floor. The support-by-fire position shift fires to the lower floor of the same structure (Figure 2-27).

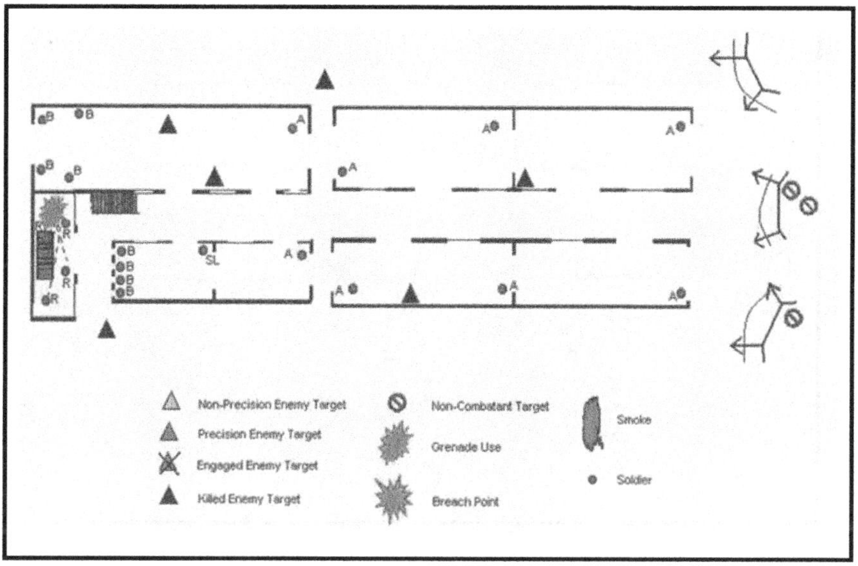

Figure 2-27. Team B, second squad--moving into the supporting fire position.

2-66. Once the floor is secure, the assault team moves down the stairs and clears the remaining floor (Figure 2-28). Meanwhile, the platoon repositions, and then consolidates and reorganizes.

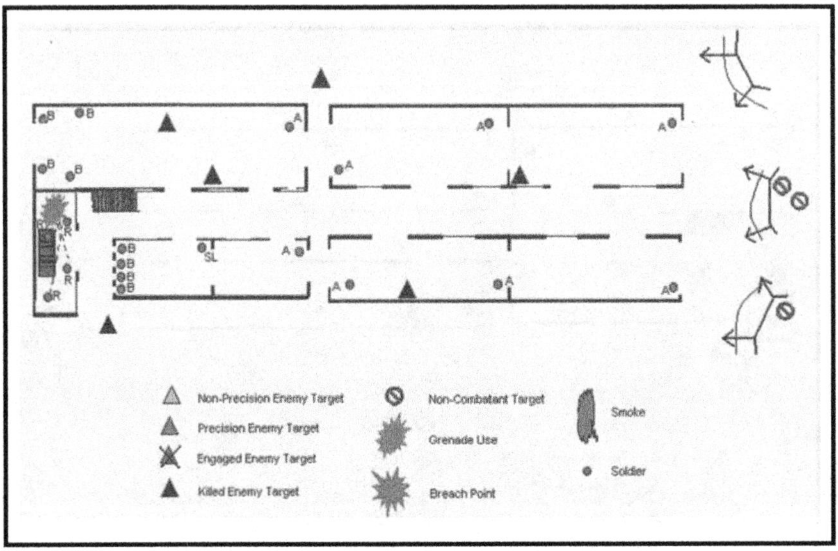

Figure 2-28. Consolidation and reorganization.

STATION 3: GRENADIER GUNNERY TRAINER

2-67. This is a live-fire station where M203 dual-purpose weapon gunners master the engagement of point-type M203 and M16 targets at 50 to 150 meters in an urban area. Tactical movement and fire commands are also trained at this station.

DESCRIPTION

2-68. This station has two main parts:

Wooden Facade

2-69. The first part is a wooden facade wall that looks like a two-story building. It is about 20 feet high by 30 feet wide. It has a doorway with outside steps, two lower floor windows and three upper floor windows, and two basement windows at ground level (Figure 2-29).

Target Engagement Lane and Firing Positions

2-70. The second part consists of the target engagement lane and seven firing positions, which are between the wooden façade and the first firing position. These positions have cover such as sandbags, log walls, and rubble. Firing points can be adjusted to support a variety of training scenarios (Figure 2-30).

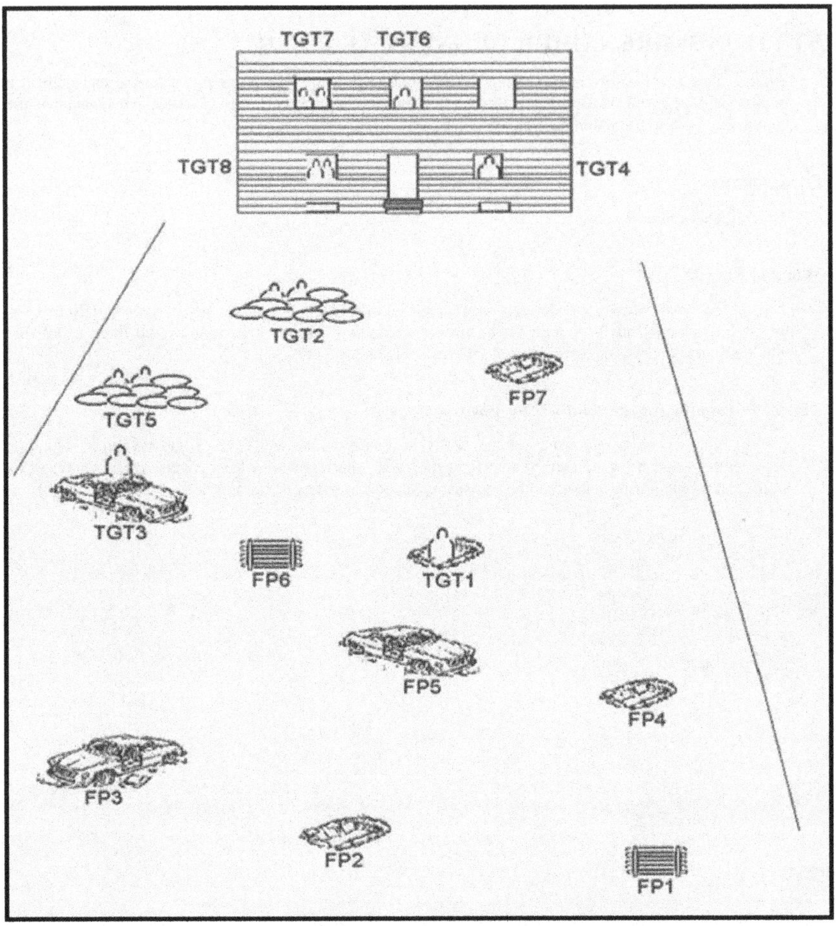

Figure 2-29. Part 1, Grenadier Gunnery Trainer, Station 3, wooden facade.

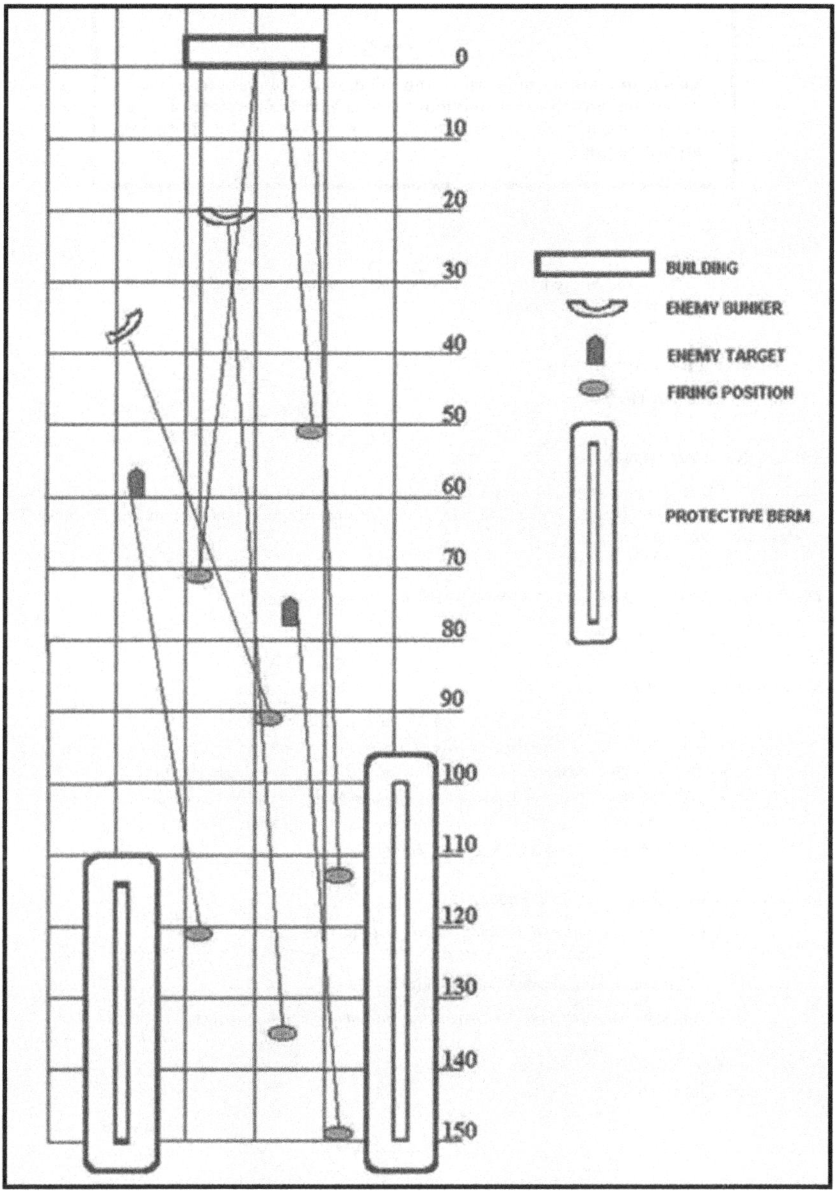

Figure 2-30. Part 2, Grenadier Gunnery Trainer, Station 3, target engagement lane and firing positions.

WARNING

Never engage targets with the 40-mm at ranges less than 31 meters, which is the minimum safe range. Trainers ensure the station accommodates this requirement as well as the 14-meter arming range.

TARGETRY

2-71. This station has twelve target mechanisms. These mechanisms are arranged so as to influence the gunner's decision as to which munition to use.

CONTROLLER REQUIREMENTS

2-72. As required by unit.

SAFETY REQUIREMENTS

2-73. Check the unit SOP and local safety regulations regarding live-fire ranges. Ensure all weapon systems are on SAFE before moving to the next firing position. Ensure that no 40-mm target is engaged at less than minimum arming range.

INDIVIDUAL AND COLLECTIVE TASKS TRAINED

2-74. These lists are noninclusive:

Supporting Individual Tasks

- 071-311-2007, *Engage Targets with M16-Series Rifle/M4 Carbine Rifle.*
- 071-311-2130, *Engage Targets with M203 Grenade Launcher.*
- 071-326-0503, *Move Over, Through, or Around Obstacles (except Minefields).*
- 071-326-0541, *Perform Movement Techniques during an Urban Operation.*
- 071-326-0557, *Select Hasty Firing Positions during an Urban Operation.*
- 071-326-0608, *Use Visual Signaling Techniques.*

Related Tactics, Techniques and Procedures

- FM 3-06.11, *Combined Arms Operations in Urban Terrain.*
  -- Chapter 3, *Offensive Operations.*
  -- Chapter 5, Fundamental Combat Skills.
- Appendix M, *Field Expedient Breaching of Common Urban Barriers.*

Training Support Requirements

2-75. Table 2-5 shows the minimum materials required to train at this station.

Table 2-5. Support requirements for Station 3.

Ammunition	*Rounds for each Weapon*	*Targets Required*
5.56-mm Ball	8	Four nonprecision human target mechanisms.
40-mm TP	8	Eight nonprecision human target mechanisms.

Conduct of Training

2-76. Whereas Table 2-5 showed the minimum amount of ammunition required for this station, Table 2-6 shows an example sequence of events and times, and Table 2-7 shows an example target scenario.

Table 2-6. Sequence of events and times for Station 3.

Event	*Action*	*Estimated Time*
1	*Perform Movement Techniques* *Select Hasty Firing Position* *Engage Targets With 5.56-mm (dry run)* *Engage targets with 40mm*	20 min
2	*Perform Movement Techniques* *Select Hasty Firing Position* *Engage Targets with 5.56mm (live-fire)* *Engage targets with 40mm*	10 min
NOTE: Training should include techniques outlined in FM 3-06.11		

Table 2-7. Example target scenario (presentation order) for Station 3.

Firing Position	*Weapon*	*Presentation Time*	*Range to Target*	*Target Presented*
1	5.56mm	6 sec	67 meters	Single Human Target (TGT 1)
2	40mm	2 min	105 meters	Multiple Human Targets (TGT 2)
3	5.56mm	6 sec	63 meters	Single Human Target (TGT 3)
4	5.56mm	6 sec	105 meters	Single Human Target (TGT 4)
5	40mm	2 min	57 meters	Multiple Human Targets (TGT 5)
6	40mm	2 min	70 meters	Multiple Human Targets (TGT 6)
6	5.56mm	6 sec	70 meters	Single Human Target (TGT 8)
7	40mm	2 min	50 meters	Multiple Human Targets (TGT 7)

STATION 4: URBAN OFFENSE/DEFENSE BUILDING, TASK/TECHNIQUE TRAINER

2-77. This station trains to attack or defend a building. Collective and individual TTP are trained. The platoon practices the task steps and performance measures for offensive and defensive tasks. This station increases the complexity of command, control, and maneuver (Figure 2-31).

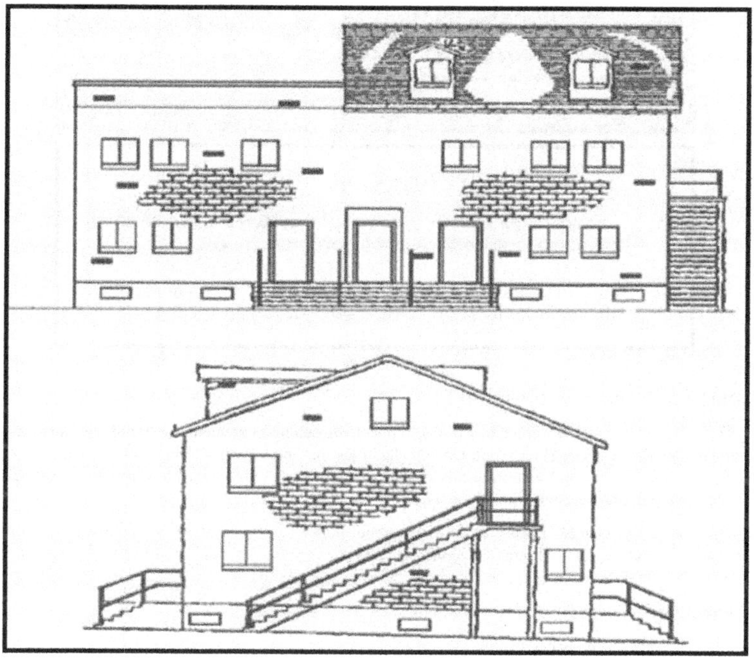

Figure 2-31. Urban offense/defense building, Station 4.

DESCRIPTION

2-78. This station is a two-story building with a basement. Each floor has several rooms, windows, doorways, loopholes, and mouseholes. The gabled roof has windows and a hatch to the attic below and half of the rooftop is flat. The building has exterior and interior stairways.

TARGETRY

2-79. This station has ten target mechanisms.

CONTROLLER REQUIREMENTS

2-80. These are based on unit requirements.

SAFETY REQUIREMENTS

2-81. Only smoke, practice, and M84 stun grenades may be used at this station. It is not a live-fire station. Training is conducted using TES with blank 5.56-mm, or TES or SESAMS. Units should consult with their supporting combat engineers for the proper demolition effects simulators (DES) to use against the blow panels throughout the station.

DANGER

TO AVOID ASPHYXIATION, NEVER DETONATE SMOKE GRENADES INSIDE BUILDINGS. THEY ARE *ALL* DANGEROUS IN AN ENCLOSED SPACE.

WARNINGS

1. **To avoid injury, place your weapon on SAFE when climbing through obstacles such as windows, fences, and walls.**
2. **To avoid inflicting carbon burns or more serious injury, never fire blank ammunition at other Soldiers within 1 meter.**

DEMOLITION EFFECTS SIMULATORS

2-82. Demolition effects simulators (DES) replicate breaching to gain entry to buildings and rooms. The walls of the UAC, Shoot House, and CACTF facilities may be breached using wall breach charge DES. Generally, the interior and exterior walls in these facilities are made with blow panels, plasterboard, or similar material, and measure about 4 feet square.

2-83. The DES charges for use at the UAC, shoot house, and the CACTF are constructed to breach a predetermined entry point without damaging the facility.

Field-Expedient Charges

- Flex-linear shaped charge.
- Doorknob charge.
- E-silhouette charge.

Operational Charges

2-84. Unit trainers must consult with installation safety and combat engineers to determine the appropriate reduced charge. See also FM 3-06.11.

INDIVIDUAL OR COLLECTIVE TASKS TRAINED

Collective Task

- FM 3-21.8 (7-8).
  - -- 07-3-9018, *Clear a Building (Platoon/Squad).*
  - -- 07-6-1216, *Defend an Urban Area (Battalion/BCT).*

Related Collective Tasks

- 07-2-5027, *Conduct Consolidation and Reorganization (Company/Platoon).*
- 08-2-0003. *Treat Casualties.*
- 08-2-0004. *Evacuate Casualties.*
- 19-3-3100, *Process Detainees at a Collecting Point (CP).*
- 07-2-1342, *Conduct Tactical Movement (Company/Platoon).*
- 07-2-1189, *Conduct a Dismounted Tactical Road March (Company/Platoon).*
- 07-2-1198, *Conduct a Mounted Tactical Road March (Company/Platoon).*
- Warrior Battle Drill 12, *Enter and Clear a Room.*

Supporting Individual Tasks

- 052-193-1013, *Neutralize Booby Traps.*
- 071-010-0006, *Engage Targets with M249 Machine Gun.*
- 071-311-2007, *Engage Targets with M16-Series Rifle/M4 Carbine Rifle.*
- 071-315-2308, *Engage Targets with M16-Series Rifle/M4 Carbine Using a Night Vision Sight AN/PVS-4.*
- 071-325-4407, *Employ Hand Grenades.*
- 071-326-0501, *Move as a Member of a Fire Team.*
- 071-326-0503, *Move Over, Through, or Around Obstacles (except Minefields).*
- 071-326-0541, *Perform Movement Techniques during an Urban Operation.*
- 071-326-0557, *Select Hasty Firing Positions during an Urban Operation.*
- 071-326-0608, *Use Visual Signaling Techniques.*
- 071-326-5605, *Control Movement of a Fire Team.*
- 071-326-5611, *Conduct the Maneuver of a Squad.*
- 071-710-0008, *Operate Night Vision Goggles AN/PVS-7B.*
- 113-571-1022, *Perform Voice Communications.*
- 191-377-5250, *Handle Enemy Personnel and Equipment.*

Related Tactics, Techniques, and Procedures

- FM 3-06.11, *Combined Arms Operations in Urban Terrain.*
  -- Chapter 3, *Offensive Operations.*
  -- Chapter 5, *Fundamental Combat Skills.*
  -- Appendix M, *Field Expedient Breaching of Common Urban Barriers*
- FM 3-22.9, *Rifle Marksmanship.*

Training Support Requirements

2-85. Table 2-8 shows the minimum ammunition required to train at this station.

Table 2-8. Ammunition requirements for Station 4.

Ammunition	*Rounds for each Weapon*
5.56-mm Blank	16
5.55-mm Blank (Linked)	24
7.62-mm Blank	100
Practice Grenade Bodies	5
Practice Grenade Fuzes	10 per platoon
M84 Stun Grenade Bodies	5
M84 Stun Grenade Fuzes	10 per platoon
Smoke Grenades	4 per platoon

Conduct of Training

2-86. Table 2-9 shows the sequence of training for Station 4. Figures 2-32 and 2-33 show example scenarios of training events that can be conducted at this station. These are only examples. The unit should develop training events based on its METL and training needs.

Table 2-9. Sequence of events and times for Station 4.

Event	*Action*	*Estimated Time*
1	*Clear a Building*	2 Hours Crawl
		1 Hour Walk
		30 Minutes Run
2	*Defend Built-up Area/Building*	1 Hours Crawl
		1 Hour Walk
		2. Hour Run
3	*Conduct AAR*	1 Hour

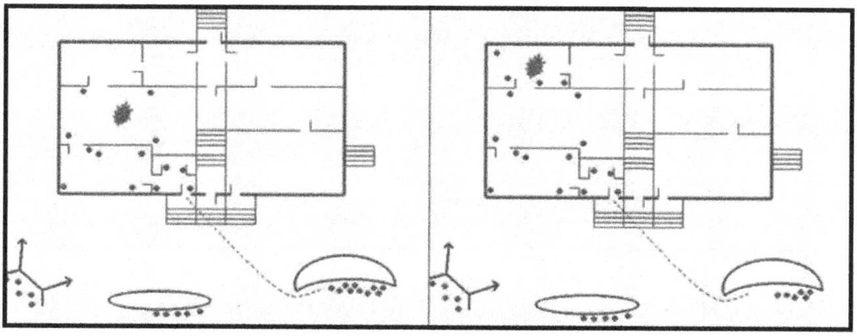

Figure 2-32. Event 1, *Clear a Building.*

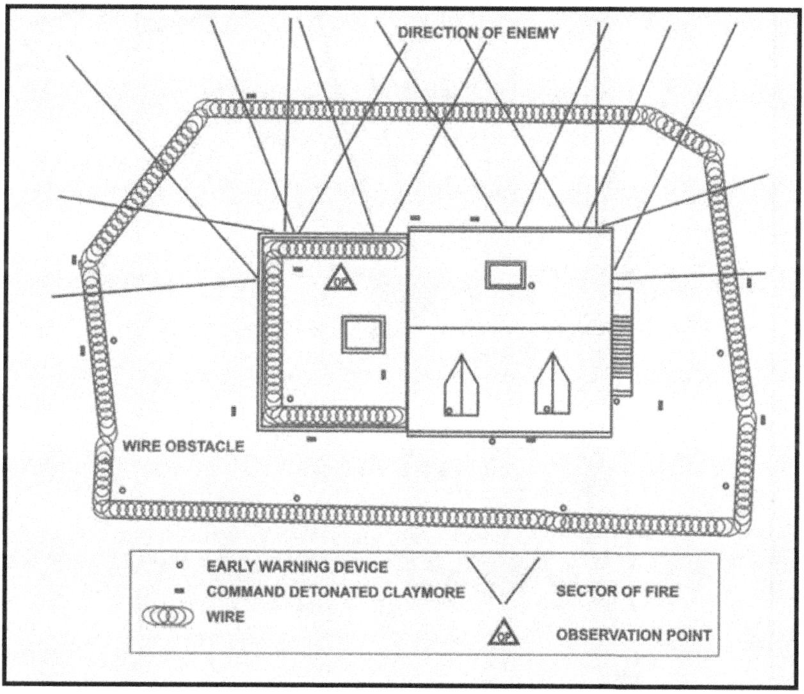

Figure 2-33. Event 2, *Defend a Building.*

STATION 5: UNDERGROUND TRAINER

2-87. This station provides training up to squad level for subterranean operations, specifically, the task *Clear and Move Tactically in a Subterranean Environment.*

DESCRIPTION

2-88. This station is a sewer system built in an irregular pattern, with four manhole covers. Two access tunnels lead from the outside tunnel to the underground loop, and the station is completely covered by dirt, only the manholes and ground level entrances are exposed (Figure 2-34).

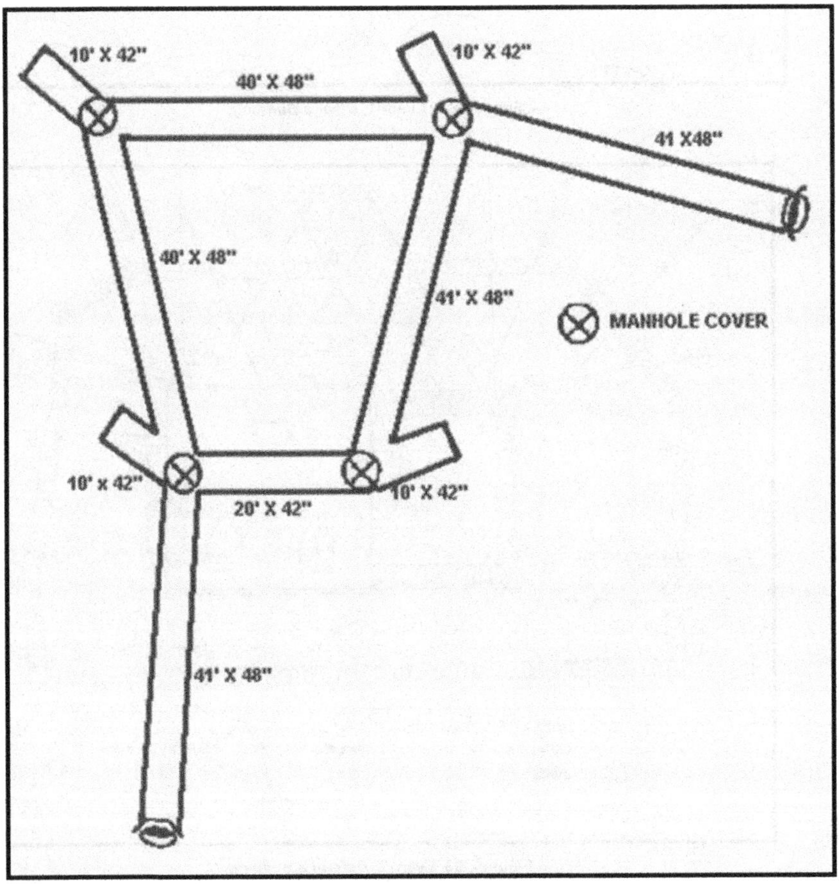

Figure 2-34. Underground Trainer, Station 5.

TARGETRY

2-89. Station 5 is a force-on-force training station and has no targetry.

CONTROLLER REQUIREMENTS

2-90. As required by unit.

SAFETY REQUIREMENTS

2-91. Soldiers should wear their helmets at all times when in the sewer system to avoid head injuries. Smoke grenades or M84 stun grenades *will not* be used in the underground trainer.

> **DANGER**
>
> TO AVOID ASPHYXIATION WHEN IN THE TRAINER, NEVER DETONATE SMOKE GRENADES IN THE SEWER. USING ANY TYPE OF EXPLOSIVE SIMULATOR IS DANGEROUS IN AN ENCLOSED SPACE. THE PROTECTIVE MASK IS NOT EFFECTIVE IN OXYGEN-DEFICIENT ATMOSPHERES.

INDIVIDUAL OR COLLECTIVE TASKS TRAINED

Collective Task

- *Move Tactically in a Subterranean Environment.*
- *Reconnoiter Area.*

Related Collective Tasks

- 07-2-1270, *Conduct an Infiltration (Company/Platoon).*
- 08-2-0003. *Treat Casualties.*
- 08-2-0004. *Evacuate Casualties.*

Supporting Individual Tasks

- 052-193-1013, *Neutralize Booby Traps.*
- 071-010-0006, *Engage Targets with M249 Machine Gun.*
- 071-311-2007, *Engage Targets with M16-Series Rifle/M4 Carbine Rifle.*
- 071-315-2308, *Engage Targets with M16-Series Rifle/M4 Carbine Using a Night Vision Sight AN/PVS-4.*
- 071-326-0501, *Move as a Member of a Fire Team.*
- 071-326-0503, *Move Over, Through, or Around Obstacles (except Minefields).*
- 071-326-0541, *Perform Movement Techniques during an Urban Operation.*
- 071-326-0557, *Select Hasty Firing Positions during an Urban Operation.*
- 071-326-0608, *Use Visual Signaling Techniques.*
- 071-326-5605, *Control Movement of a Fire Team.*
- 071-326-5611, *Conduct the Maneuver of a Squad.*
- 071-710-0008, *Operate Night Vision Goggles AN/PVS-7B.*

- 113-571-1022, *Perform Voice Communications.*
- 191-377-5250, *Handle Enemy Personnel and Equipment.*

Related Tactics, Techniques and Procedure

- FM 3-06.11, *Combined Arms Operations in Urban Terrain.*
  -- Chapter 3, *Offensive Operations.*
  -- Chapter 5, *Fundamental Combat Skills.*
  -- Appendix M, Field Expedient Breaching of Common Urban Barriers.
- FM 3-22.9, *Rifle Marksmanship.*

TRAINING SUPPORT REQUIREMENTS

2-92. Use Table 2-10 for the minimum ammunition required to train at this station.

Table 2-10. Ammunition requirements for Station 5.

Ammunition	*Rounds for each Weapon*
5.56-mm Blank	20
5.55-mm Blank (Linked)	40

CONDUCT OF TRAINING

2-93. Table 2-11 shows the sequence of training for this station. These are only examples. The unit should develop training events based on its METL and training needs.

Table 2-11. Sequence of events and times for Station 5.

Event	*Action*	*Estimated Time*
1	*Reconnoiter*	30 minutes (Crawl) 30 minutes (Walk) 30 minutes (Run)
2	*Conduct an Infiltration (Co/Plt)*	30minutes (Crawl) 30 minutes (Walk) 30 minutes (Run)
3	*Conduct AAR*	30 minutes

Chapter 3

Shoot House

This chapter describes the live-fire Shoot House and provides layouts and training scenarios. Fighting in built-up areas is fragmented; gains are often small. Therefore, units depend on the initiative, skill, and discipline of small-unit leaders and individual Soldiers. Leaders must be competent and confident in their urban operations skills and must have the courage to accomplish their missions while isolated from their parent units. Soldiers and leaders require mission-type orders that allow for decentralized execution.

PURPOSE

3-1. The Shoot House provides the leader with a facility to train and evaluate the unit during a live-fire exercise. Units are trained and evaluated on their ability to conduct tactical movement, engage targets, conduct breaches, and practice target discrimination in an urban environment.

DESCRIPTION

3-2. This facility has two parts, the live-fire facility and the AAR room:

LIVE-FIRE FACILITY

3-1. The live-fire facility is a single-level structure with a barn roof, catwalk, and eight adjoining rooms. Each room has video capability for AARs. The walls are bulletproof and prevent ricochets. Each side has an entrance. Replaceable blow panels throughout the facility can withstand reduced DES blasts. These also offer alternative entry points. A crane system in the roof allows trainers to selectively replace doorways and blow panels with solid wall sections. Before training explosive and mechanical breaching operations, trainers should place mobile exterior baffles or containment walls placed by each entry to prevent rounds and demolition effects from escaping. All of these capabilities let the unit vary its training scenarios. For safety, the Shoot House is windowless (Figure 3-1).

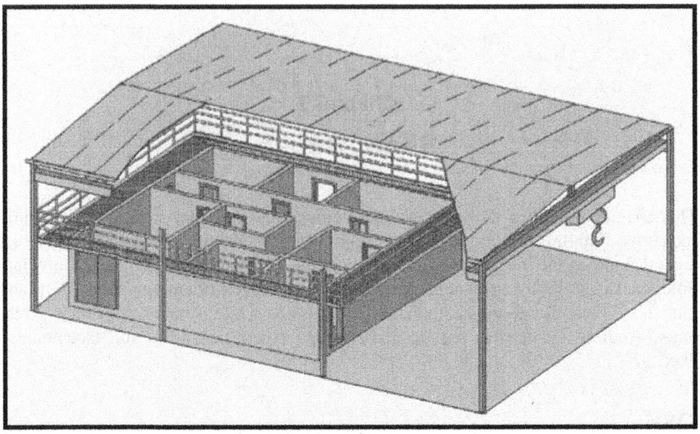

Figure 3-1. Shoot House.

AAR BUILDING

3-2. The AAR building has a conference/AAR room that can accommodate up to a platoon size element, an instrumentation and targetry control room, and an AAR edit and projection room that can handle a formal AAR (Figure 3-2).

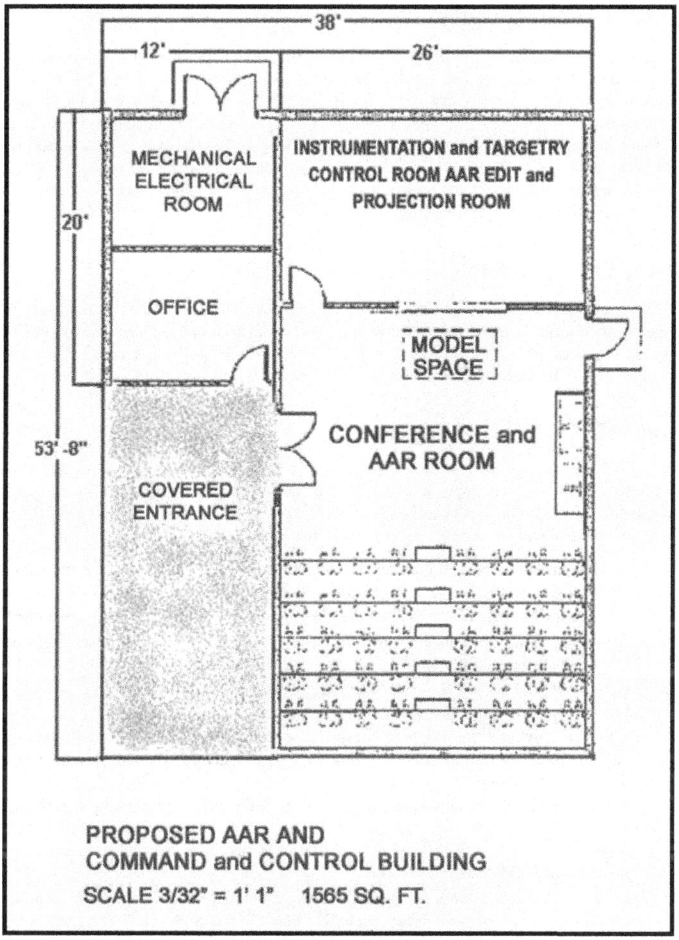

Figure 3-2. Command and control building and AAR room.

TRAINING AUDIENCE

3-3. The facility provides leaders the ability to train and evaluate the unit during a live-fire exercise. However, the unit's training requirements determine the size of the unit and the tasks to be trained at this facility. The unit develops its own organization and the steps for conducting the exercise based on its METL and training needs.

TRAINING INFORMATION

3-4. Tactical operations in an urban environment may range from combat missions to stability and support missions. Commanders should note that the Shoot House *is not* suited for the employment of live-fragmentation/concussion grenades, 40-mm HE/HEDP, antiarmor weapons, mortars, artillery, or air delivered munitions. Crew-served weapon systems, M2 and M1 tanks, and machine guns can be employed against supplemental targetry that is exterior to the Shoot House. The Shoot House has specially constructed areas that are used to conduct either mechanical or explosive breaching techniques. When conducting explosive breaching, units should consult with local Engineers for appropriate DES charges and consult Appendix A before conducting the training.

CONDUCT OF TRAINING

3-5. The unit develops realistic time frames for each operation or task. The times are based on the *crawl-walk-run* concept. The training event or STX can be arranged in any order as long as they follow a tactical scenario. *Assault a Building*, *Enter a Building and Clear a Room*, and *Clear a Building* are the primary collective tasks trained at this facility.

EXECUTION

3-6. Tables 3-1 through 3-5, pages 3-4 through 3-7, shows a sequence of events and times, and Figure 3-3 through Figure 3-6, pages 3-6 through 3-9, show example scenarios. These are only examples. The unit should develop training events based on its METL and training needs. However, the leader should do the following when conducting training at the live-fire Shoot House:

1. Identify the unit's critical missions (see commander's training guidance), and select those tasks that support specific tasks related to urban operations.
2. Arrange the related tasks in the order of their importance to mission accomplishment.
3. Assess the unit's current level of proficiency (strengths and weaknesses), and determine the degree of difficulty of each task. The training events and scenarios should start from the least difficult to the most difficult tasks. This gives the Soldiers confidence.
4. Identify the individual and leader tasks to be trained.
5. Conduct individual training first, because it is the first step in the building-block approach. It is also the basis for unit or collective task training. Mastery of individual skills is a precondition for effective training.
6. Set up training conditions based on unit proficiency. At first, keep the conditions basic, until Soldiers attain proficiency. Increase difficulty and realism as training progresses. Then, add limited visibility, CBRN factors, and live fires with service or short-range training ammunition (SRTA).
7. Understand that the Shoot House is a live-fire capable training facility. However, it can be used to train any level of proficiency, including blank, TES, SESAM, SRTA, and ball (no tracers). The leader assesses the unit's proficiency.

Table 3-1. Sequence of events and times for the platoon.

Element	*Event action*	*Estimated Time*
All 1st Squad 2d Squad 3rd Squad	1. Conduct precombat checks (for example; install, align, and troubleshoot MILES equipment; conduct inspection; draw equipment). 2. Support Operations. 3. Conduct Rehearsals. 4. Conduct Event 1 (Crawl).	2 Hours
3rd Squad 1st Squad 2d Squad	5. Support Operations. 6. Conduct Rehearsals. 7. Conduct Event 1 (Crawl).	2 Hours
2d Squad 3rd Squad 1st Squad	8. Support Operations. 9. Conduct Rehearsals. 10. Conduct Event 1 (Crawl).	2 Hours
1st Squad 2d Squad 3rd Squad	11. Support Operations. 12. Conduct Rehearsals. 13. Conduct Event 1 (Walk).	2 Hours
3rd Squad 1st Squad 2d Squad	14. Support Operations. 15. Conduct Rehearsals. 16. Conduct Event 1 (walk).	2 Hours
2d Squad 3rd Squad 1st Squad	17. Support Operations. 18. Conduct Rehearsals. 19. Conduct Event 1 (walk).	2 Hours
1st Squad 2d Squad 3rd Squad	20. Support Operations. 21. Conduct Rehearsals. 22. Conduct Event 1 (Run).	1 Hours
1st Squad 2d Squad 3rd Squad	23. Support Operations. 24. Conduct Rehearsals. 25. Conduct Event 1 (Run).	1 Hours
1st Squad 2d Squad	26. Support Operations. 27. Conduct Event 1 (Run).	1 Hours

Table 3-2. Event 1, Squad Sequence (*Crawl*).

Element	*Event Action*	*Estimated Time*
1 (CRAWL)	**Terminal Learning Objective** Familiarize the squad with how to enter and clear a room during the crawl phase of training **Tasks** *Enter and Clear a Room (Squad)* • *Perform Movement Technique* • *Select Hasty Firing Position* *Conduct a Breach (Mechanical)* • *Employ Hand Grenades* • *Engage Targets with M16A2/M4* • *Engage Targets with M249 Machine Gun*	**2 Hours**
NOTES: 1. Perform controlled pairs and target discrimination techniques (FM 3-22.9). 2. Include techniques (FM 3-06.11) and performance measures (FM 3-21.8 and DTMS).		

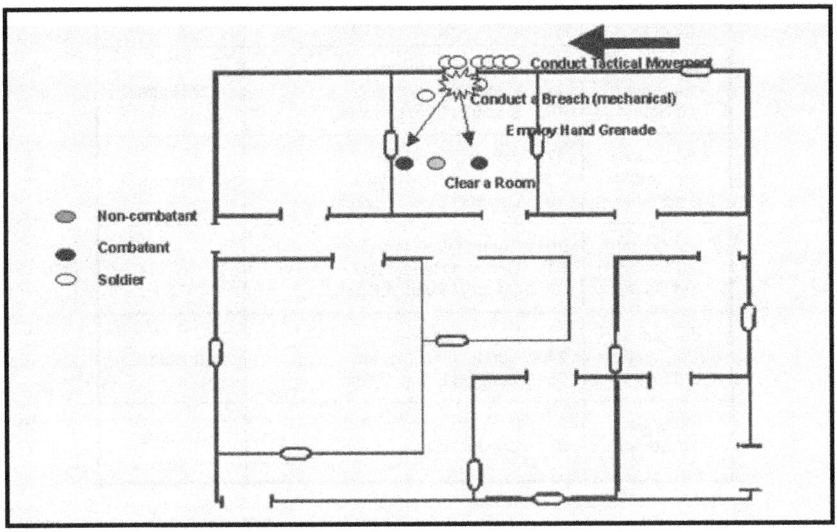

Figure 3-3. Event 1, *Squad Scenario (Crawl)*, force-on-targetry.

Table 3-3. Event 2, Squad Sequence (*Walk*).

Element	*Event Action*	*Estimated Time*
2 (WALK)	**Terminal Learning Objective** Familiarize the squad with how to enter and clear multiple rooms during the walk phase of training **Tasks** *Enter and Clear a Building (Squad)* • *Perform Movement Technique* • *Select Hasty Firing Position* *Conduct a Breach (Explosive DES)* • *Engage Targets with M16A2/M4* • *Engage Targets with M249 Machine Gun* *Conduct AAR*	**2 Hours**
NOTES: 1. Perform controlled pairs and target discrimination techniques (FM 3-22.9) **2. Include techniques (FM 3-06.11) and performance measures (FM 3-21.8 and DTMS).**		

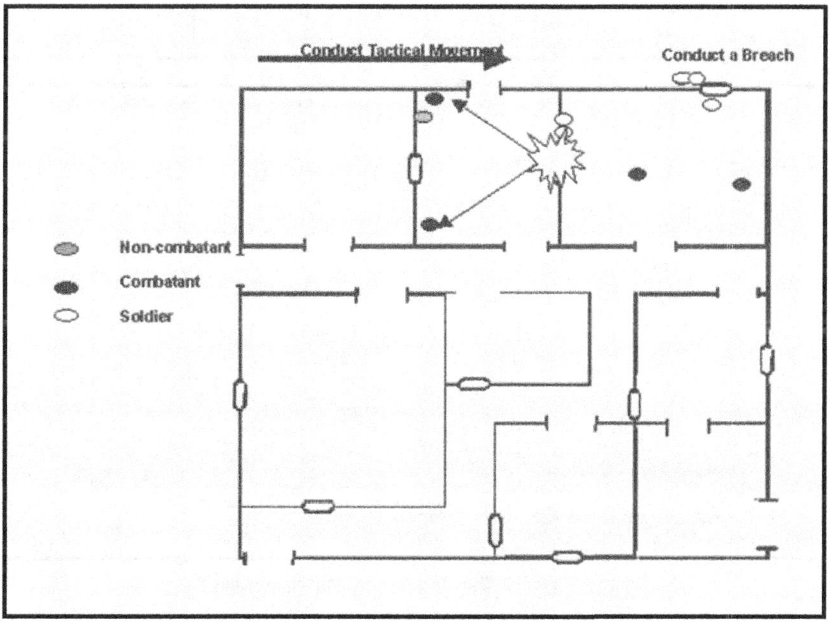

Figure 3-4. Event 2, *Squad Scenario (Walk)*, force-on-force.

Table 3-4. Event 3, Squad Sequence (*Run*).

Element	*Event Action*	*Estimated Time*
3 (RUN)	**Terminal Learning Objective** Refamiliarize squad with how to enter and clear multiple rooms during the run phase of training. **Tasks** *Enter and Clear a Building (Squad)* • *Perform Movement Technique* • *Select Hasty Firing Position* *Conduct a Breach (Mechanical)* • *Engage Targets with M16A2/M4* • *Engage Targets with M249 Machine Gun* *Clear a Hallway* • *Engage Targets with M16A2/M4* • *Engage Targets with M249 Machine Gun* *Conduct a Breach (explosive DES)* • *Engage Targets with M16A2/M4* • *Engage Targets with M249 Machine Gun* *Conduct AAR*	**1 Hour**

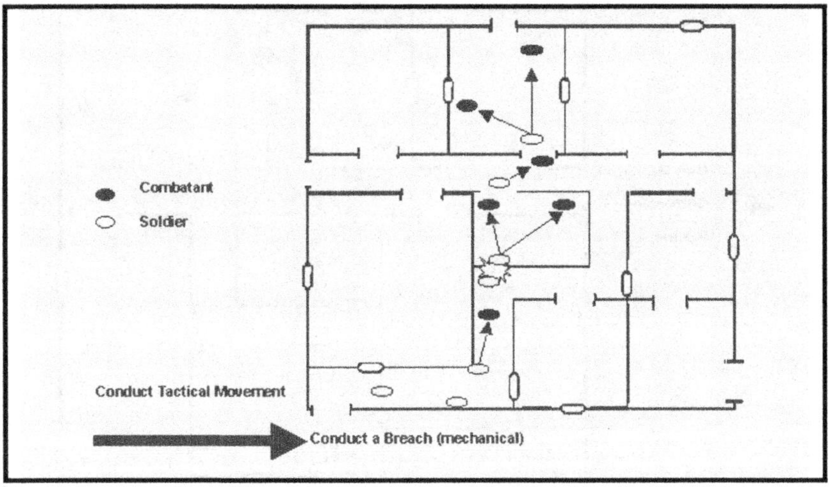

Figure 3-5. Event 3, *Squad Scenario (Run)*, force-on-targetry.

Table 3-5. Event 4, Platoon Evaluation Sequence (*Run*).

Element	*Event Action*	*Estimated Time*
4 (RUN)	*Occupy Assembly Area*	1 Hour
	Develop and Communicate a Plan	2 Hours
	Prepare for Combat	30 min
	Perform Passage of Line	1 Hour
	Conduct Tactical Movement	
	Break Contact	5 min
	Conduct Intermediate AAR	20 min
	Conduct Tactical Movement	
	Perform Overwatch/Support by Fire	1 Hours
	Assault a Building	1 Hours
	Consolidate and Reorganize	2 Hours
	Conduct Final AAR	1 Hours

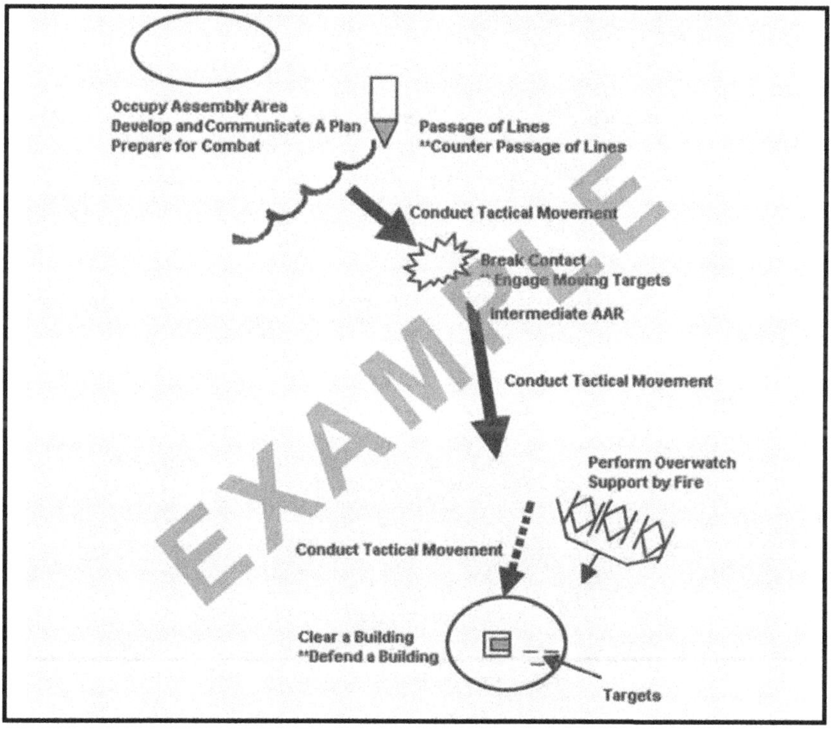

Figure 3-6. Example platoon tactical scenario (*Run*).

OPERATION ORDER

3-7. Figure 3-7 shows an example OPORD for the conduct of training.

____
(Classification)

Copy ____ of ___ copies
____Company _____ Infantry
(Unit Location)____________
Date/time Group __________

(Message Reference Number)

OPORD: _____________
Reference Map, Series ___________________
Time Zone used Throughout the Order:_____________________

TASK ORGANIZATION:

______Platoon Platoon
______Platoon ____ 60-mm Section
______Antiarmor Section

1. **SITUATION**

a. **Enemy Forces**. Annex B (Intelligence Overlay [prepared by battalion S-2]).

(1) The company is opposed by elements of the Red Company of the Blue Brigade at about 60 percent strength. Within our zone, we have a platoon-sized element occupying a defensive position in a building in the vicinity of (______). The enemy has small arms weapons and may call on company mortars for support.

(2) Current indications are the enemy is regrouping to initiate small-unit offensive actions in the area. His most likely course of action is to move to the (direction) to link up with elements of the parent unit.

b. **Friendly Forces**.

(1) (___) Infantry attacks at (______) hours to seize Objective ALPHA (grid). The intent is to prevent the enemy element from moving (direction) to link up with other enemy units.

(2) Company (__) on our (left, right) makes the main attack at (____) hours to seize Objective RED (grid) in order to establish blocking positions to stop the enemy's movement to the (direction).

____
(Classification)

Figure 3-7. Example company OPORD.

(Classification)

2. MISSION

Company (__) attacks to seizes Objective BLUE (grid) at (date/time) hours in order to prevent the enemy from regrouping and engaging Company (__) (main effort) from Objective BLUE.

3. EXECUTION

a. **Concept of Operation**. Annex C (Operations Overlay). (This should include those steps listed in the task , Conduct Troop-Leading Procedures, paragraph 17. These should be written out when possible and not just placed on an overlay.)

(1) Maneuver. (___) Platoon establishes a defense in its zone, NLT (date/time) hours, to destroy small enemy elements moving or exfiltrating to the (direction), from the vicinity of Objective BLUE. This defense supports the attack on Objective BLUE by completing destruction of any enemy that may escape. (Evaluated) Platoon conducts the company main attack at (hours) to seize Objective BLUE to destroy the enemy or force them to withdraw into (___) Platoon's defense. (___) Platoon supports the attack of the (Evaluated) Platoon on Objective BLUE to suppress enemy fires and isolate the objective. On seizure of Objective BLUE, (Evaluated) Platoon and (___) Platoon will defend in sector to destroy the enemy elements withdrawing to avoid (___) Company's attack.

(2) **Fire support**.

(a) Priority of fires.

- Initially to (Evaluated) Platoon.
- Upon seizure of Objective BLUE, priority shifts to (Defensive) Platoon.

(b) Priority targets.

- 60-mm priority target initially AA4031.
- On initiation of assault on Objective BLUE, priority shifts to target AA4027 to isolate the objective.
- Upon seizure of Objective BLUE, priority shifts to target AA4025.

(Classification)

Figure 3-7. Example company OPORD (continued).

(Classification)

b. **Tasks to maneuver units.**

(1) (___) Platoon. Destroy enemy elements moving to the (direction), NLT (date/time).

(2) (Evaluated) Platoon. Consolidate and defend in sector to destroy enemy elements following the seizure of Objective BLUE.

(3) (___) Platoon. Be prepared to assault Objective BLUE, on order, consolidate and defend in sector to destroy enemy elements following the seizure of Objective BLUE.

c. **Tasks to Subordinate Units.**

(1) 60-mm Section. Initially in the vicinity of CP 7, priority of fires to the (Evaluated) Platoon. Move with the (Supporting) Platoon. On order, priority of fires to the (Defensive) Platoon.

(2) Antiarmor Section. Attached to the (Supporting) Platoon.

d. **Coordinating Instructions**:

(1) (Defensive) Platoon reports defense established and locations.

(2) Platoon reports the seizure of all objectives.

(3) Platoon reports the consolidation and establishment of the defense.

(4) Order of march from the assembly area to the objective: defense platoon, evaluated platoon, headquarters, support platoon, 60-mm mortar.

4. **SERVICE SUPPORT**

a. **General**. Company trains collocated with the battalion combat trains in the vicinity of (grid). Supply distribution at the company CP.

(Classification)

Figure 3-7. Example company OPORD (continued).

(Classification)

b. **Materiel and Services.**

(1) Class I: Ration cycle MRE-MRE-MRE.

(2) Class V: Pick up basic load at company CP, NLT (date/time).

(3) Medical: Company casualty collection point is __________.

(4) EPW: EPW collection point is ___________.

5. **COMMAND AND SIGNAL**

a. **Command.** Company command group initially follows the (Evaluated) Platoon.

b. **Signal.**

(1) Radio-listening silence is in effect until contact is made.

(2) SOI _____ is in effect.

JONES
Commanding

AnnexES:
B-Intelligence Overlay
C-Operations Overlay
D-Fire Support Overlay

(Classification)

Figure 3-7. Example company OPORD (continued).

TRAINING SUPPORT REQUIREMENTS

3-8. Minimum trainers and evaluators should be based on unit proficiency, training needs, and local safety requirements. Table 3-6 contains consolidated support requirements.

Table 3-6. Consolidated support requirements.

Ammunition	*Event 1*	*Event 2*	*Event 3*	*Event 4 (Optional)*
5.56-mm (per weapon) ball, blank, SRTA, or SESAMS	20 Blank	20 SESAMS	20 Ball/SRTA	20 Ball/SRTA
5.56-mm linked (per weapon)	30 Blank	30 SESAMS	30 Ball/SRTA	30 Ball/SRTA
Hand grenade body (practice)	0	--	4	4
Hand grenade fuzes (practice) per platoon	0	--	4	4
Hand grenade, smoke, per platoon	0	6	6	6
M84 hand grenade, stun, per platoon	0	4	4	0
TES (MILES Equipment)				
M16 (1 per weapon, 1 per OPFOR)	NO	YES	NO	NO
M249 (1 per weapon)	NO	YES	NO	NO
SAAF (1 per platoon)	NO	YES	NO	NO
Controller Gun (1 per OC)	NO	YES	NO	NO
Targetry	5	0	5	10
OPFOR	NO	6	NO	NO

TRAINING AND EVALUATION

3-9. The T&EOs are the foundation of the MTP and the collective training of the platoon. T&EOs are training objectives (task, conditions, and standards) for the collective tasks that support platoon critical wartime operations. T&EOs may be trained separately, in an STX, in an FTX, or in live-fire exercises. For collective live-fire standards, the trainer needs to refer to the applicable Infantry Gunnery Manual for the appropriate course of fire. Those standards and courses of fire need to be integrated into the training exercise. Table 3-8 provides a list of related collective and supporting individual tasks.

Table 3-7. Related collective and supporting individual tasks.

Task	FM 3-21.8	FM 3-06.11	STP 21-1-SMCT	STP 7-11B1-SM-TB	STP 21-24-SMCT
Conduct Consolidation and Reorganization Treat Casualties Evacuate Casualties Process Enemy Prisoners of War/Captured Materiel Defend MOUT Building Conduct Tactical Movement	X	X			
Supporting Individual Tasks					
Engage Targets with an M16-Series Rifle			071-311-2007		
Engage Targets with an M16-Series Rifle Using a Night Vision Sight AN/PVS-4			071-315-2007		
Engage Targets with an M249 Machine Gun				071-010-0006	
Employ Hand Grenades			071-325-4407		
Move Over, Through, or Around Obstacles (Except Minefields)			071-326-0503		
Neutralize Booby Traps					052-193-1013
Move as a Member of a Fire Team				071-326-0501	
Perform Movement Techniques during MOUT		X		071-326-0541	
Conduct the Maneuver of a Squad				071-326-5611	
Conduct an Explosive Breach				071-440-0030	
Execute a Mechanical Breach				071-440-0027	
Conduct a Breach				071-440-0029	
Operate Night Vision Goggles AN/PVS-7				071-710-0008	
Select Hasty Firing Positions during MOUT				071-326-0557	
Control Movement of a Fire Team				071-326-5605	

Chapter 4

Breach Facility

Urban operations require Soldiers to enter buildings through walls, doors, windows, and roofs. The Breach Facility is designed to train Soldiers using tasks and techniques on how to breach locked doors, windows, and create man-sized holes in walls. The facility can be used to train mechanical, ballistic, thermal, and explosive breaching.

PURPOSE

4-1. The Breach Facility (Figure 4-1) is used to train Soldiers semiannually on the technical aspects of breaching techniques. It is also used to train TTP and explosive techniques not trained on any other type range.

Figure 4-1. Breach facility.

DESCRIPTION

4-2. This facility has three stations:

STATION 1

4-3. This station is a facade that is 8 feet high and 88 feet wide with four sections of two doors for a total of eight doors. Soldiers can conduct mechanical, thermal, ballistic, and explosive breaching techniques on doors at this station. Half of the doors open from the exterior, and the other half of the doors open from the interior (Figure 4-2).

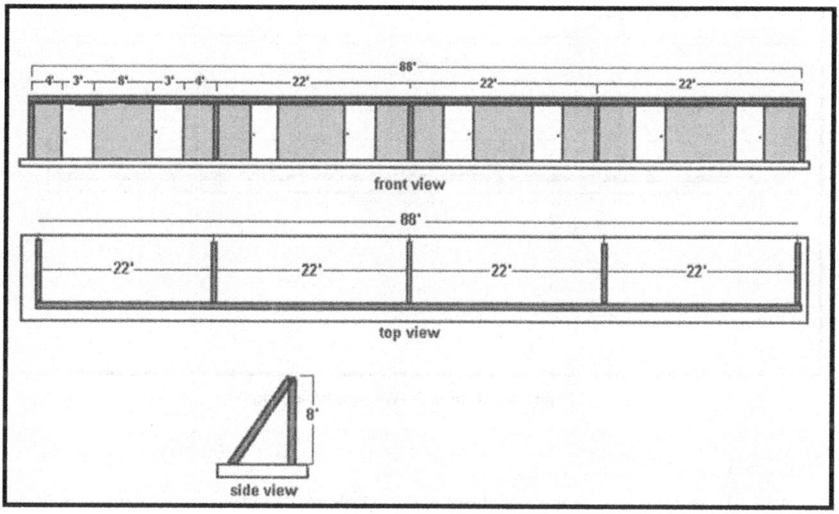

Figure 4-2. Station 1—Door breaching.

STATION 2

4-4. This station is a facade. It measures 10 feet high by 88 feet wide. It has four sections of two windows each, for a total of eight windows. Soldiers can conduct mechanical and explosive breaching techniques on windows at this station (Figure 4-3).

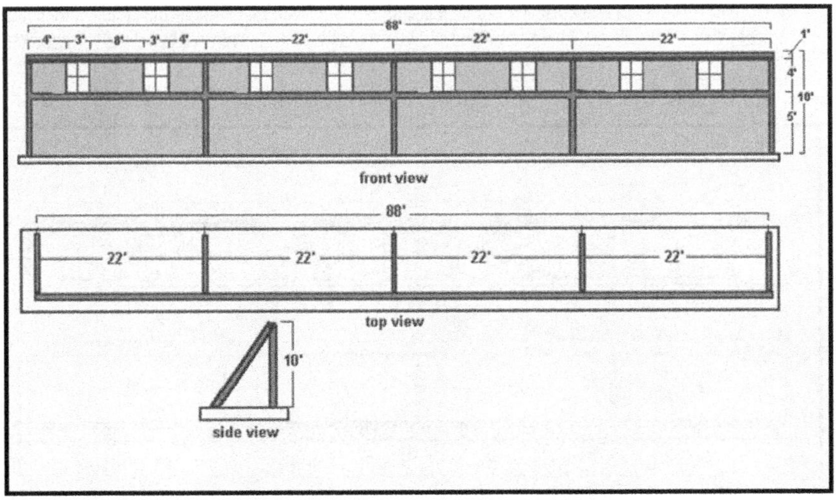

Figure 4-3. Station 2—Window breaching.

STATION 3

4-5. This station is a facade that looks like a wall of a building. It measures about 8 feet high by 28 feet wide. Precast panels will fit into each of the three sections. The panels measure 8 feet square and can be made of concrete, concrete blocks, stone, brick, wood or other material. The panels are designed for use with thermal and explosive breaching techniques (Figure 4-4).

TRAINING AUDIENCE

4-6. The facility provides Infantry, Engineer, MP, and Cavalry leaders with a facility to train and evaluate individual and leader breaching tasks. Unit training requirements determine the size of the unit and the tasks to be trained at this facility (Appendix B provides a unit training matrix). The individual tasks trained at the Breach Facility follow:

- 052-193-1003, *Prime Explosives Nonelectrically*.
- 071-440-0027, *Execute a Mechanical Breach*.
- 051-193-1055, *Construct a Nonelectric Initiating Assembly with Modernized Demolition Initiators (MDI)*.
- 052-193-1312, *Construct Demolition Initiating Sets*.
- *Conduct a Thermal Breach* (Engineers only).

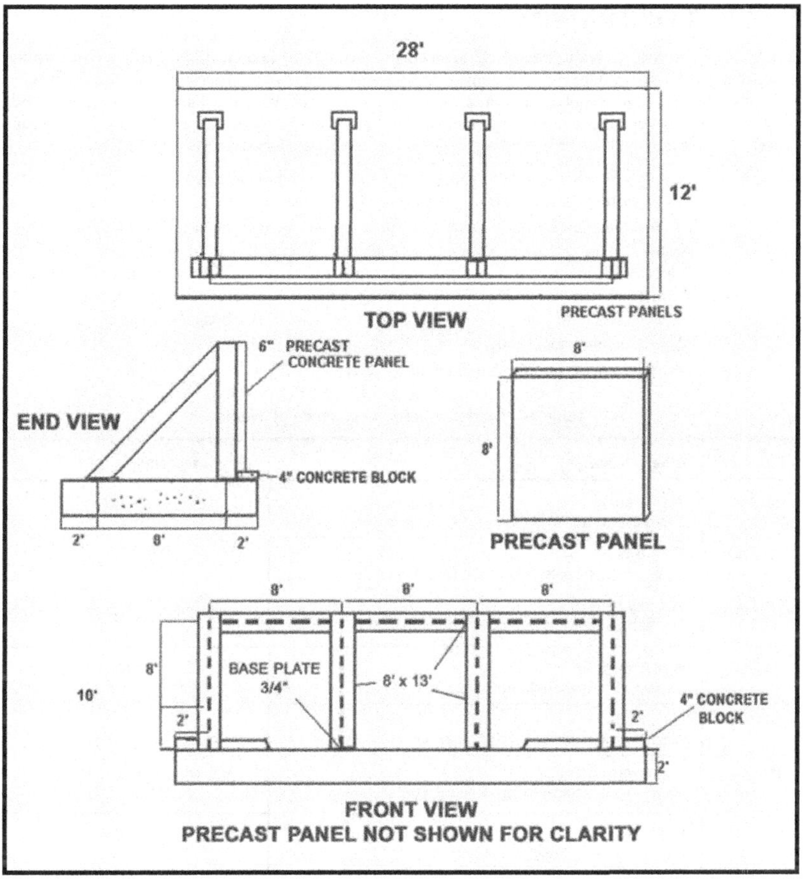

Figure 4-4. Station 3—Wall breaching.

TRAINING INFORMATION

4-7. Many types of organizations can train at this facility. Tactical operations in an urban environment may range from combat missions to stability and support actions. Commanders should note that this facility is an individual task and technique trainer. Training at the Breach Facility should be conducted prior to conducting collective training at the UAC, shoot house, and the CACTF. Trainers *must be certified* before conducting demolition training per local SOP or ACOM policy.

CONDUCT OF TRAINING

4-8. Trainers develop realistic time requirements for each task to be trained based on the *crawl-walk-run* concept and the size of the unit to be trained. The training event may be arranged in any order as long as it builds from simple to complex task and techniques. Table 4-1 contains an example sequence of events to assist units in developing their training plan. Leaders should do the following prior to conducting training. Before conducting ballistic breaching with automatic weapons, leaders should consult FM 3-06.11:

- Identify the unit's critical missions (see commander's training guidance) and select those tasks that support specific METL related to urban operations.
- Arrange the related tasks in the order of their importance to mission accomplishment.
- Assess the unit's current level of proficiency (strengths and weaknesses) and determine the degree of difficulty of each task. The training events should start from the less difficult tasks to the more difficult tasks. This instills confidence in the Soldiers conducting the training. Leaders should consult FM 3-06.11 for more information on explosive breach techniques.
- Identify the individual and leader tasks that are to be trained.
- Set up training conditions based on unit proficiency.

Table 4-1. Example platoon sequence of events.

Element	*Event Action*	*Estimated Time*
3d Sqd	Wall Breach: Explosive (crawl, walk, run)	3 hours
1st Sqd	Door Breach: Mechanical, ballistic, explosive (crawl, walk, run)	
2d Sqd	Window Breach: Mechanical, explosive (crawl, walk, run)	
2d Sqd	Wall Breach: Explosive (crawl, walk, run)	3 hours
3d Sqd	Door Breach: Mechanical, ballistic, explosive (crawl, walk, run)	
1st Sqd	Window Breach: Mechanical, explosive (crawl, walk, run)	
1st Sqd	Wall Breach: Explosive (crawl, walk, run)	3 hours NOTE: Walls, doors, and windows require replacement after each run/iteration to train to standards.
2d Sqd	Door Breach: Mechanical, ballistic, explosive (crawl, walk, run)	
3d Sqd	Window Breach: Mechanical, explosive (crawl, walk, run)	

TRAINING SUPPORT REQUIREMENTS

4-9. Table 4-2 shows minimum requirements based on three breach teams for each platoon-sized element. This facility is resource intensive, and Table 4-3 shows the total annual requirements per platoon-sized element.

Table 4-2. Consolidated minimum support requirements.

Ammunition/Equipment	*Crawl*	*Walk*	*Run*
Mechanical Breach: *Prying , Striking, or Cutting Tool*			
Door	(talk through)	1 per team (defeat one lock)	1 per team (defeat hinges)
Window	(talk through)	1 per team (defeat one lock)	1 per team (defeat hinges)
Ballistic Breach: Door			
Shotgun [1]	(talk through)	2 rds per weapon (defeat one lock)	5 rds per weapon (defeat hinges)
GREM (inert) [2]	(talk through)	1 per team – inert	1 per team - service
[1] If available to unit. [2] Grenade Rifle Entry Munition (GREM).			
Explosive Breach:			
Door: Det Cord (Flex Linear Charge) Nonelectric Firing System	(build)	(practice emplacement)	29 feet per team 1 per team
Wall: C4 (Satchel Charge) Nonelectric Firing System	(build)		4 blocks per team
		(practice emplacement)	23 feet per team 1 per team
Window: Det Cord (Flex Linear Charge) Nonelectric Firing System	(build)		29 feet per team 1 per team
		(practice emplacement)	
Thermal Breach Equipment Sappers (Only)	1	1	1

Table 4-3. Consolidated annual support requirements.

Ammunition/Equipment	*Total*
Doors	6
Windows	6
Wall Panels	0
Shotgun Rounds	42
GREM	6 inert/2 service
Det Cord (feet)	486
C4 (blocks)	24
Nonelectric Firing Systems	18

TRAINING AND EVALUATION

4-10. Individual training is the first step in the building-block approach and sets the basis for unit or collective training. Mastery of individual skills is an essential precondition for urban operations training and a good method for conducting individual training is through concurrent training. Concurrent training is designed to train groups of Soldiers on different tasks or techniques and these tasks or techniques may or may not be related. For example, a leader may subdivide the unit at the Breach Facility into breaching orders. One group of Soldiers may be conducting an explosive breach, while other groups are conducting mechanical and ballistic breaches. Ballistic breaching and explosive breaching does have a SDZ, an MSD, and ricochet considerations.

SAFETY CONSIDERATIONS

4-11. Safety requires effort on the part of everyone. It is not enough to teach safety rules; the rules must be practiced and enforced. Depending on the nature of the obstacle, the breacher may use mechanical, ballistic or explosive techniques. The following guidelines should be followed when training at the Breach Facility.

- Wear all appropriate personal protective equipment and follow appropriate guidelines:
  -- Safety glasses.
  -- Hearing protection.
  -- Body armor.
  -- Helmet.
  -- Gloves.
  -- Elbow and knee pads.
- Inspect equipment before use.
- Use appropriate tool.
- Use appropriate amount of explosive not to exceed four blocks of C4. Place the C4 on the wall only, not on the foundation, as this would cause irrepairable damage.
- Stay within the surface danger zone.
- Always use the minimum safe distance.
- Do not modify tool.
- Ensure proper use and maintenance of hydraulic and pneumatic tools; tools can burst under pressure when improperly used.

Chapter 5

Combined Arms Collective Training Facility

The home station CACTF is designed to conduct multiechelon, full-spectrum operations training up to battalion TF level. Example company and platoon STX lanes, and stability and support operations tasks, are provided to help units develop training plans. Leaders use training exercises to train, evaluate, and practice performance-oriented collective tasks. TEWT, STX, and FTX are the three types of exercises covered in this chapter. Each is designed to help the trainer develop urban operations training skills and increase unit proficiency in an urban environment. The urban operations collective training exercises were developed using the standard CACTF as described in this chapter and may be modified to fit available urban operations facilities.

PURPOSE

5-1. The leader or commander is provided with a CACTF to evaluate unit urban operations proficiency. The CACTF is intended to support blank fire, MILES/TES, SESAMS, STX, and FTX scenarios on a semiannual basis. The CACTF supports branch-specific lane training and combined arms training up to battalion level across the full spectrum of operations: offense, defense, stability, and support.

DESCRIPTION

5-2. The CACTF replicates an urban environment. The facility consists of 2.25 square kilometers of urban sprawl with 20 to 26 buildings, roads, alleys, parking areas, underground sewers, parks, athletic fields, and command and control building. The actual size and configuration of the CACTF depends on the local installation site requirements. The CACTF is designed to support heavy and light Infantry, Armor, artillery, and aviation positioning and maneuver. Table 5-1 shows the recommended types of buildings and features in the CACTF.

Table 5-1. CACTF characteristics.

20 to 26 Buildings (1.5 km by 1.5 km)	
Tunnel and sewer system	Props and furniture
Shantytown	Targetry
One 3-story building	Audio and image capture with EDIT/REPLAY capability
Three 2-story buildings	Control building with AAR facility
Industrial area	Breachable walls
Electricity and potable water	
City dump	

5-3. The buildings have one, two, or three stories (some with basements), sloped and flat roofs. CACTF building variations pose different tactical and technical training challenges. Figure 5-1 shows an example CACTF containing 24 buildings with the following characteristics:

- The hotel is a dominating structure typical of a central business district. The hotel has an elevator shaft, fire escapes, and a large first-floor lobby.
- The two-story office building contains a series of identical rooms.
- The school has a long, central corridor, large windows, and a single large room.
- The townhouse has multiple sections with a common attic.
- The service station accommodates combat vehicles in its service bay.

5-4. The buildings have other features that enhance the CACTF's training value.

- Constructed mouse-holes permit movement between rooms and floors.
- Loopholes in roofs and outside walls allow observation and fields of fire.
- Roof hatchways lead to the top floors of multiple story buildings.

5-5. Rooms vary in size, and in number of doorways and windows. These variations require Soldiers to identify rooms that provide suitable indoor fighting positions for TOW, Javelin, Dragon, and AT4 antitank weapon systems.

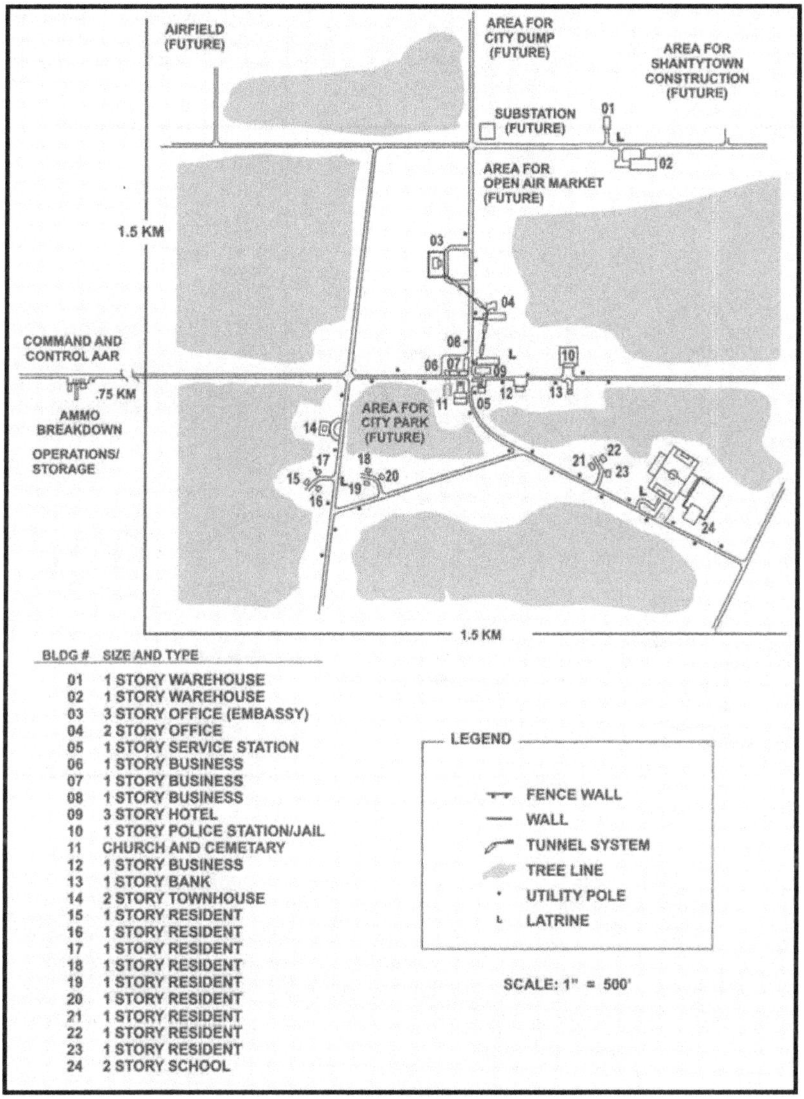

Figure 5-1. Example CACTF schematic.

5-6. Buildings are arranged in a realistic urban pattern. The CACTF represents an area consisting mainly of residential, commercial, public institutional, and light industrial buildings. Buildings and streets are arrayed with sufficient dispersion to allow up to nine platoon-sized STX sectors (Figure 5-2) or six company/team STX sectors (Figure 5-3).

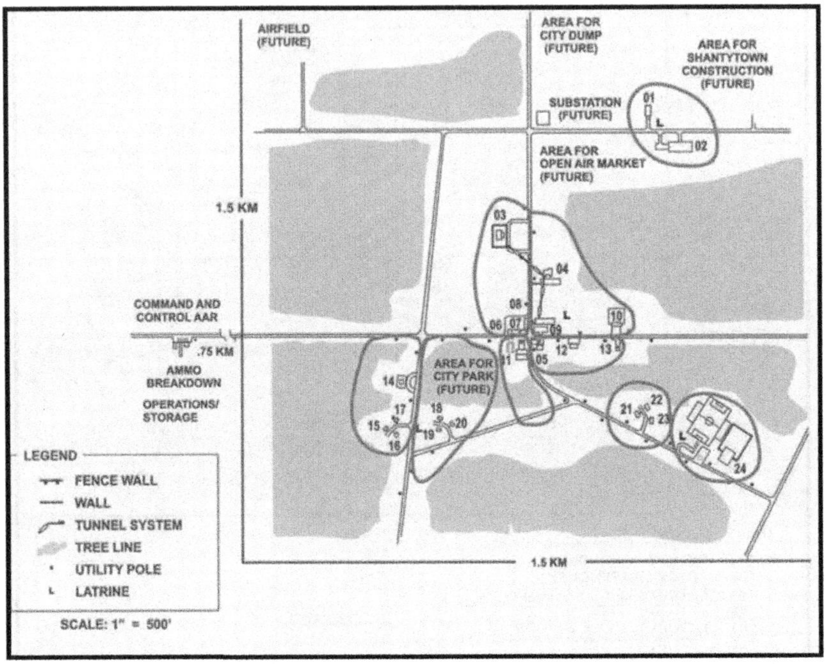

Figure 5-2. Platoon ODSS STX sectors.

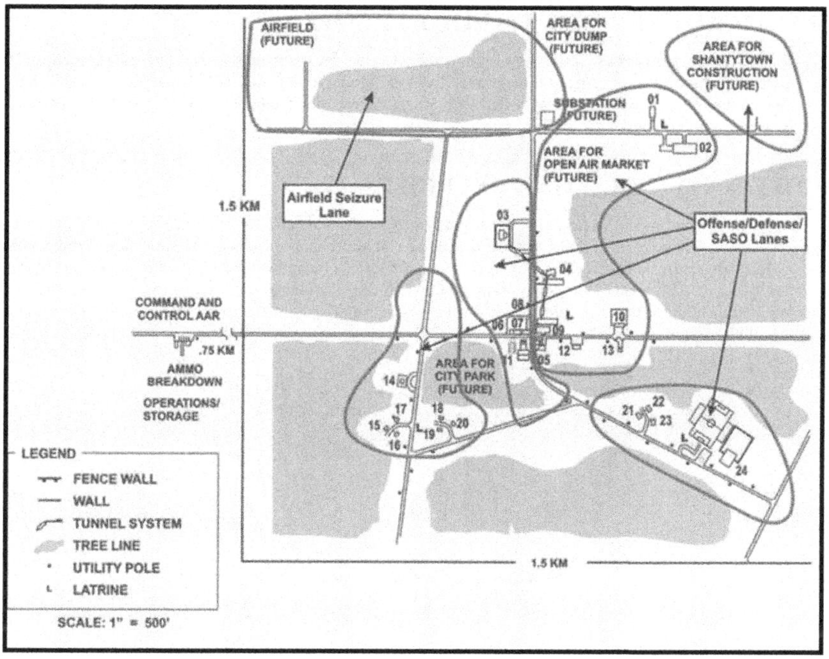

Figure 5-3. Company/team STX sectors.

5-7. Other urban features pose a variety of terrain considerations. The street network includes one and two-lane primary, secondary, local, and service roads. They join in T-shaped, L-shaped, and four-way intersections. Other features may be used to enhance realism such as signs, benches, furniture, newsstands, utility poles, street lights, vegetation, simulated rubble, live civilians on the battlefield, role players, and OPFOR. Open areas next to the CACTF can be used for maneuvering and staging areas. The specific details and layout of the CACTF may vary from one installation to another.

CAUTION

Trainers must ensure that fighting position reinforcement does not exceed the weight limits of the buildings in order to avoid structural damage and safety hazards to the Soldiers.

TACTICAL EXERCISE WITHOUT TROOPS

5-8. A terrain walk and a TEWT may be conducted in a local downtown area to prepare leaders and to take advantage of the available time in the CACTF. Terrain walks and TEWTs are conducted to teach leaders the tactical value of urban terrain. Appendix A provides training tips and additional information regarding the conduct of a TEWT.

COLLECTIVE TRAINING EXERCISES

5-9. The following exercises were developed IAW the guidance in FM 7-0, FM 7-1, FM 25-4, and TC 7-98-1. These publications include offensive and defensive scenarios for combat in urban areas, specifically, two platoon STXs and two company FTXs. Each can be conducted with blank ammunition; with MILES/TES; with 9-mm or 5.56-mm SESAMS for short-range precision target engagements (less than 50 meters); or force on force. Each was chosen to help the commander develop, sustain, and evaluate his unit's mission proficiency:

- Infantry Rifle Platoon Attack STX
- Mechanized Infantry Platoon Defend STX
- Infantry Rifle Company Attack FTX
- Mechanized Infantry Company Team Defend FTX

INFANTRY RIFLE PLATOON STX ATTACK

1. **Objective.**

This example STX trains collective, leader, and individual tasks for the platoon operation, Attack in a Built-Up Area.

2. **Interface.**

This STX is supported by two warrior drills, *React to Contact* and *Enter and Clear a Building*.

3. **Training.**

a. *Guidance*. FM 7-0 and FM 7-1 provide training guidance, tips, and enhancers for planning and executing STXs. The trainer should review the individual, leader, and collective tasks to be performed during the STX to determine the ones that require initial or refresher training.

b. *General Situation*.

(1) The exercise scenario follows: Contact with the enemy has been reestablished. Initial reports indicate he is at 65 to 70 percent strength, and that he has not received reinforcements. His defensive positions are in the CACTF and are not well established. He has indirect fire and CAS; he has already used chemicals, and will probably do so again. The platoon receives an order to prevent the enemy from establishing a heavily fortified defense in the CACTF. Such a defense could prevent friendly forces from continuing offensive operations.

(2) This exercise begins when the platoon leader receives the company FRAGO, and it ends after the platoon consolidates and reorganizes on the objective. An AAR should be held after completing the task, *React to Contact*, and at the completion of the exercise. If necessary, portions of the exercise should be repeated until the platoon performs them to standard. Figure 5-4 provides a graphic scenario of task performance for the STX *Attack*. Table 5-2 provides a recommended sequence for performance of T&EOs, and the time allotted for each portion of the STX.

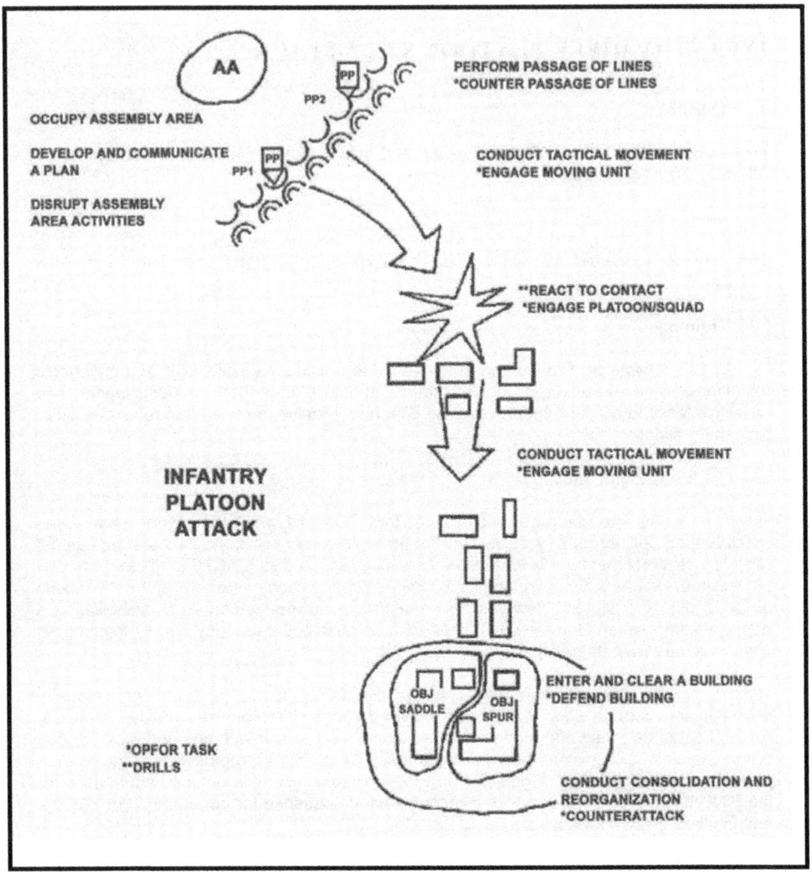

Figure 5-4. Attack STX scenario.

Table 5-2. Sequence of events and times for STX Attack.

Event	*Action*	*Estimated Time*
1	*Occupy Assembly Area*	1 hour
2	*Develop and Communicate a Plan*	1 hour
3	*Perform Passage of Lines*	1 hour
4	*Conduct Tactical Movement*	1 hour
5	*React to Contact (Drill)*	30 min
6	*Conduct Intermediate AAR*	30 min
7	*Conduct Tactical Movement*	30 min
8	*Enter and Clear a Building (Drill)*	1 hour
9	*Conduct Consolidation and Reorganization*	1 hour
10	*Conduct Final AAR*	1 hour
	* **Total Time**	**8 hours, 30 min**
* Conducting an exercise at night or in increased MOPP requires additional time.		

c. *Special Situation.* The platoon is part of a company in a secure assembly area when a FRAGO to attack is received (Figure 5-5).

1. **SITUATION.** The company is opposed by one platoon.

a. *Enemy Forces.* The enemy is at 65 to 70 percent strength. He is preparing a defense in CACTF (**MOUT site**) to prevent friendly forces from continuing to attack along the CACTF avenue of approach. He is expected to use artillery-delivered nonpersistent nerve agents in the defense of CACTF.

b. *Friendly Forces.* (Battalion designation) attacks (**date-time group**) to destroy enemy forces at OBJ COWBOY (**CACTF**) in order to pass follow-on forces, allowing friendly forces to continue the attack.

2. **MISSION.** (___) Company seizes OBJ HORSE (CACTF) vicinity (**grid**) NLT (**date-time group**) to prevent the enemy from establishing a heavily fortified defense.

3. **EXECUTION.**

a. *Concept of the Operation.* (See overlay.). The company moves in a wedge to Assault Position 1, assault CACTF to seize OBJ HORSE, consolidate, and reorganize.

(1) Maneuver. (**Evaluated**) Platoon will lead the company and be the main effort in the attack. They seize OBJ SADDLE vicinity (**grid**) to allow passage of friendly elements through CACTF. (___) Platoon moves on the company's left and seize OBJ SPUR vicinity (**grid**) to protect the left flank of the main effort. (___) Platoon moves on the company's right and establish a support-by-fire position located vicinity (**grid**).

(2) Fire support. fires are used to obscure movement to OBJ HORSE and to suppress the enemy during the assault. Priority of indirect fires is to (*evaluated*) Platoon, which is allocated one priority target (**120-mm mortar**).

Figure 5-5. Example FRAGO for STX Attack.

```
 b. Tasks to Maneuver Unit.
    (1)  (Evaluated) Platoon.
       (a)  Use Passage Point 1.
       (b)  Orient defense from TRP 2 to TRP 3.
       (c)  Be prepared to employ a wire-mined obstacle vicinity TRP 5
during consolidation.
    (2)  (___) Platoon.
       (a)  Use Passage Point 2.
       (b)  Orient defense from TRP 1 to TRP 2.
    (3)  (___) Platoon
       (a)  Use Passage Point 2.
       (b)  Orient defense from TRP 3 to TRP 4.
 c. Coordinating Instructions.
    (1)  Company RP is located (grid).
    (2)  Company linkup is (grid).
    (3)  MOPP1 is in effect (date and time).
    (4)  Selective rubbling is authorized. Destruction of an area
larger than one city block requires brigade commander approval.
```

Figure 5-5. Example FRAGO for STX Attack (continued).

4. **Support Requirements.**

a. *Minimum Trainers/Evaluators*: This exercise can be conducted with the company commander or platoon leader acting as the trainer or primary evaluator. At least one more controller or evaluator is required with the OPFOR. Other platoons being trained or evaluated will increase support personnel requirements.

b. *Vehicles/Communications*: All vehicles organic to the platoon should be included. The OPFOR should also have some combat vehicles.

c. *Opposing Force*:. The OPFOR should consist of at least a reinforced squad.

d. *Maneuver Area*: A training area at least 4 by 15 km is needed for cross-country movement and infiltration, and the CACTF is the ideal objective area. The terrain should offer multiple covered and concealed approaches to the objective.

5. **Evaluation.**

Table 5-3 lists the T&EOs from DTMS used to evaluate this STX.

Table 5-3. T&EOs used to evaluate STX Attack.

Task/Function	*Task Number*
Perform Passage of Lines	7-3-1125
Conduct Tactical Movement	7-2-1342
Occupy Assembly Area	7-3-1136
Develop and Communicate a Plan	7-3-1605
Conduct Consolidation and Reorganization (Company/Platoon)	7-2-5027

MECHANIZED INFANTRY PLATOON STX DEFEND

1. **Objective**. This example STX trains collective, leader, and individual tasks for the platoon operation, *Defend Urban Area/Building*.

2. **Interface.** This STX is supported by two warrior drills: *React to Contact* and *Enter and Clear a Building*.

3. **Training.**

a. *Guidance*. FM 7-0 and FM 7-1 provide training guidance, tips, and enhancers for planning and executing STXs. The trainer should review the individual, leader, and collective tasks to be performed during the STX to determine which require initial or refresher training.

b. *General Situation.*

(1) The exercise scenario is as follows: Latest INTSUMs indicate that enemy forces are expected to use a high-speed avenue of approach to counterattack. The approach is dominated by a small urban area containing strongly constructed buildings. The enemy has the capability for indirect fire, CAS, and chemicals. A defense in and around the town is necessary to deny the enemy the use of the avenue of approach. The platoon, as part of a larger force, is ordered to occupy and prepare defensive positions that overlook the avenue of approach. Indirect fire is available.

(2) Conduct of the exercise begins when the platoon receives an OPORD to move to establish a platoon defensive position. An AAR should be conducted after Soldiers enter a building and clear a room, and a final AAR should be conducted once all evaluation notes are compiled. If necessary, portions of the exercise should be repeated until the platoon's performance is satisfactory.

(3) When the platoon receives a warning order to prepare to move to establish a defense, the platoon is part of a company, which is part of a task force, and is located in an assembly area. Figure 5-6 provides a graphic scenario of task performance for the Defend STX. Table 5-4 provides a recommended sequence for performance of T&EOs and the time allotted for each portion of the STX.

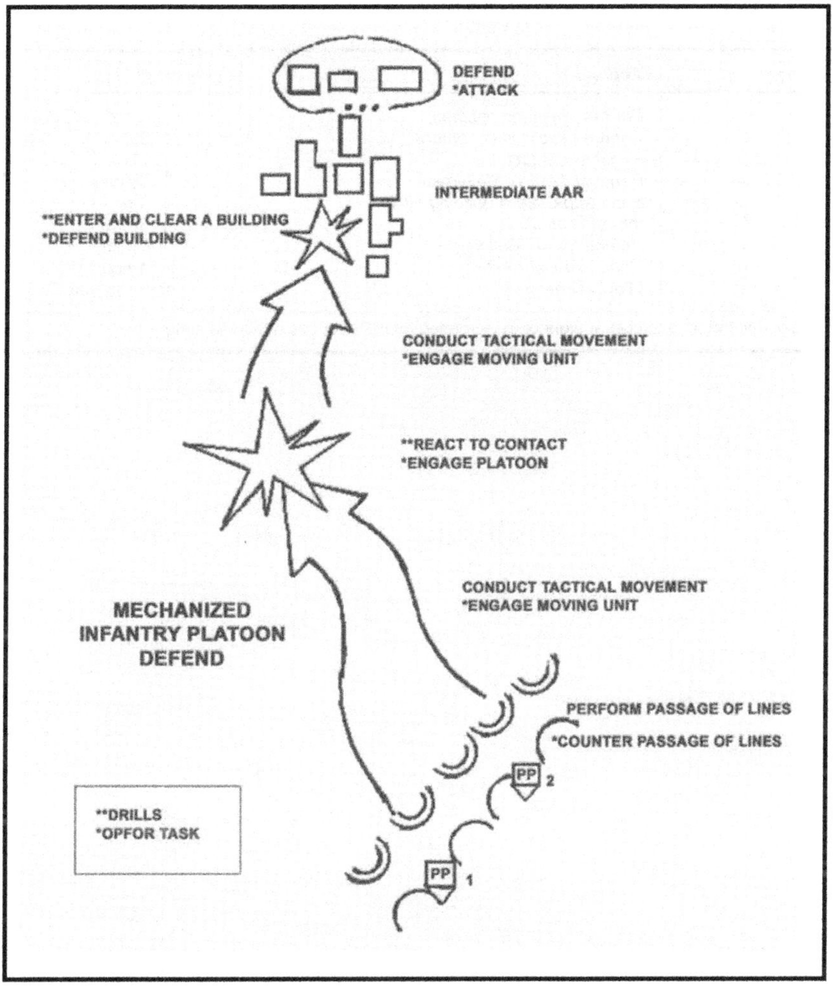

Figure 5-6. Defend STX scenario.

Table 5-4. Sequence of events and times for STX Defend.

Event	*Action*	*Estimated Time*
1	*Perform Passage of Lines*	1 hour
2	*Conduct Tactical Movement*	30 min
3	*React to Contact*	1 hours
4	*Conduct Tactical Movement*	30 min
5	*Enter and Clear a Building (Drill)*	1 hour
6	*Intermediate AAR*	30 min
7	*Defend MOUT/Building*	6 hours
8	*Conduct Final AAR*	1 hour
	***Total Time**	**11 hr 30 min**
*Conducting an exercise at night or in increased MOPP requires additional time.		

c. *Special Situation.* Thirty minutes after receiving the warning order, the platoon receives the company OPORD (Figure 5-7).

____
(Classification)

Copy no ___ of ___ copies
___ Battalion, ___ Infantry
Unit location or place of issue
Date-time group (with time zone)
Message reference number

OPERATION ORDER NO. _____
References: Map, series no. ___, sheet no. ___, edition ___
Time Zone Used Throughout the Order: ____
TASK ORGANIZATION:
___ **Platoon,** ___ **Platoon(-),** ___ **Platoon, Company Control**

Company HQ
1/1/C-52d Engr
1/2 ___,___
(Reserve)
Company Trains

1. **SITUATION.**

a. **Enemy Forces.**
(1) *Composition, Distribution, and Strength.* We are opposed by elements of the Liberation Guards Militia, Red Brigade. Several platoon-sized outposts are located within the battalion sector and consist of about 30 to 35 Soldiers per platoon. They are equipped with an assortment of Soviet small arms, including AK-47s, RPKs, RPGs, and various wheeled and tracked vehicles including some BMPs. Enemy outposts are at 60 to 70 percent strength and have indirect fire assets available.
(2) *Capabilities.* The enemy units are well entrenched in their defensive positions. The strongpoints have dedicated fire support assets and are capable of using nonpersistent nerve agents to prevent coordinated attacks. No enemy air threat exists but some heavy armored vehicles have been sighted in the area.
(3) *Probable Course of Action.* Intelligence indicates the enemy strongpoints will continue to fortify their positions until relief arrives within 24 hours in the form of a main enemy counterattack.

b. **Friendly Forces.** (__) TF defends in sector NLT (______) along PL (____) to deny enemy forces use of avenue of approach (____). (__) TF must be prepared to continue movement forward of PL (____), on order.
(1) Missions of units on left and right are (as required).

____
(Classification)

Figure 5-7. Example OPORD for STX Defend.

(Classification)

(1) (__) TF defends in sector NLT (______) from (grid) to (grid) to (grid) to contain any penetrations of the brigade sector.

(1) (___) Battery, (___) Battalion provides DS to 1st Brigade.

c. **Attachments and Detachments**. 1/1/C-52d Engineers attached.

2. **MISSION**. (___) Company blocks enemy forces NLT (date-time group) to prevent them from using high-speed avenues of approach through (name of MOUT site), and to prevent the enemy from counterattacking into the battalion sector.

3. **EXECUTION**.

a. **Concept of the Operation**. (See overlay.) Company prevents the enemy from using the high-speed avenues of approach through (MOUT site) by moving to the site and establishing a blocking position, which overwatches the platoon BPs covering the main avenues. We will block in depth by placing two platoons forward and one back, and maintaining a reserve to block any penetration. We must ensure our sectors of fire are placed to mass fire at the decisive point.

(1) *Maneuver*. (Evaluated) Platoon acts as the main effort, blocking from BP (__), vicinity (grid), to prevent enemy penetration of PL (_____). (___) Platoon occupies BP (__), vicinity (grid), to block enemy movement into the main effort's left flank. (___) Platoon occupies BP (__) to prevent the enemy from making a coordinated attack on the main effort.

(2) *Fire Support*. The purpose of fires is to disrupt the enemy when his counterattack reaches PL (_____). Priority of indirect fires is to (*evaluated*) Platoon, which is allocated two 155-mm priority targets, one of which may be used as an FPF.

(3) *Intelligence*. Priority of collection is to identify and locate enemy combat reconnaissance patrols and the enemy main effort.

(4) *Engineering*. During movement to the battle position, obstacles are used to ensure mobility. During blocking actions, the purpose of obstacles is to prevent the enemy from advancing on larger streets and to canalize him into more restrictive terrain. Priority of effort is to countermobility. Priority of support is to (*evaluated*) Platoon.

b. **Tasks to Maneuver Units**:

(1) *(Evaluated) Platoon*:

(a) Conduct passage of lines using PP 1.

(b) Lead the company movement after crossing the LD. Be prepared to react to enemy contact when (MOUT site) is reached.

(c) Enter and clear building(s) (number determined by commander).

(Classification)

Figure 5-7. Example OPORD for STX Defend (continued).

(Classification)

(d) Occupy BP (__) and orient on TRPs 2 and 3.
(e) Establish an OP vicinity (grid) to observe main avenue of approach.

(2) *(___) Platoon*:
(a) Conduct passage of lines using PP 2.
(b) Move second in order of movement.
(c) Enter and clear building(s) (number determined by commander).
(d) Occupy BP (__) and orient on TRP 1.

(3) *(___) Platoon*:
(a) Conduct passage of lines using PP 3.
(b) Move last in order of movement.
(c) Enter and clear building(s) (number determined by commander).
(d) Occupy BP (__) and orient on TRPs 4 and 5.
(e) Establish an OP vicinity (grid).

c. **Tasks to Subordinate Units**: Engineer squad travels with the company headquarters behind *(evaluated)* Platoon and supports the main effort's emplacement of obstacles.

d. **Coordinating Instructions**.
(1) Timings.
(a) Brief back ________.
(b) Company rehearsal ________.
(c) SP time ________.
(d) In position NLT (time) ________.
(2) Report all enemy contact.
(3) Report crossing all phase lines.
(4) MOPP status ____; enemy chemical attack is expected.
(5) Air defense status is YELLOW, WEAPONS HOLD.
(6) Leaders briefback the movement and defense plan to the commander at the company CP at (date-time group).
(7) Destruction of subterranean approaches requires brigade commander's approval.
(8) Selective rubbling is authorized. Destruction of more than one building requires commander's approval.
(9) Marking of rooms and buildings is IAW CO SOP.

(Classification)

Figure 5-7. Example OPORD for STX Defend (continued).

(Classification)

(10) Rules of Engagement follow:
(a) Minimize collateral damage to structures.
(b) Civilians have been evacuated. Evacuate to company CP any other civilians encountered.

4. SERVICE SUPPORT. IAW the battalion SOP.

a. **Miscellaneous**.
(1) Battalion trains located vicinity (grid).
(2) Company trains located vicinity (grid).

b. **Materiel and Services**.
(1) *Supply*.
(a) Class I. Ration Cycle M-M-M. Platoons must maintain a three-day supply in each BP.
(b) Class II. Submit requirements for special equipment ASAP. Coordinate with 1SG.
(c) Class III. Refuel all vehicles before (date-time group).
(d) Class IV. Coordinate with S-4 for delivery.
(e) Class V.
- Platoons cache one additional basic load in BP.
- Platoons submit request for additional demolitions NLT (date-time group).

(2) *Services*. Battalion decontamination site is located vicinity (grid).
(3) *Maintenance*.
(a) Priority to (*evaluated*) Platoon M2 BFVs.
(b) UMCP located vicinity (grid).

c. **Medical Evacuation and Hospitalization**.
(1) Battalion aid station located vicinity (grid).
(2) Company CCP located with company CP.

d. **Personnel**.
(1) No replacements for next 72 to 96 hours.
(2) Brigade mortuary affairs point located in BSA.
(3) EPW collection sites located at (grid) for battalion and (grid) for company.

(Classification)

Figure 5-7. Example OPORD for STX Defend (continued).

(Classification)

5. COMMAND AND SIGNAL.

a. **Command.** Initially, company CP will be traveling center of company wedge. Location in BP to be determined.

b. **Signal.**
(1) Current SOI in effect.
(2) Primary means of communication in BP is wire, FM, messenger, in that order.
(3) Radio listening silence is in effect until after crossing PL (____) or until contact is made.

(Classification)

Figure 5-7. Example OPORD for STX Defend (continued).

4. **Support Requirements**.

a. *Minimum Trainers/Evaluators*: This exercise is conducted by one company grade officer who is the trainer and primary evaluator.

b. *Vehicles/Communications*: All vehicles organic to the platoon should be included.

c. *Opposing Force*:. The OPFOR should be platoon strength.

d. *Maneuver Area*: A training area at least 2 by 10 km, with hasty fighting positions, is desirable for cross-country movement. The CACTF serves as the objective area. The terrain, ideally, should offer multiple covered and concealed routes. Using terrain that limits the leader to a geographic or predictable route prevents evaluation of the unit's ability to conduct a terrain analysis and select covered and concealed routes.

e. *Barrier Materials*. Sufficient Class IV (sandbags, lumber, and so on) must be made available to fortify positions, IAW Chapter 3, FM 3-06.11.

5. **Evaluation**.
Table 5-5 lists the T&EOs from DTMS used to evaluate this STX, and Table 5-6 lists the supporting tasks.

Table 5-5. T&EOs used to evaluate STX Defend.

Task/Function	*Task Number*
Defend MOUT/Building	7-3-1118
Perform Passage of Lines	7-3-1125
Conduct Tactical Movement (Co/Plt)	7-2-1342

INFANTRY RIFLE COMPANY FTX ATTACK

Table 5-6. Supporting tasks for company FTX.

STX C-II-1	*STX C-II-2*	*STX C-II-3*
Conduct an Infiltration (Co/Plt)	Perform Cordon and Search	Disarm Belligerents
Perform Linkup	Take Action on Contact	Handle Captured Insurgents or Belligerents
Execute an Assault (MOUT)	Conduct an Infiltration (Co/Plt)	Apprehend/Detain Noncombatants
Conduct Consolidation and Reorganization (Co/Plt)	Perform Linkup	
	Perform Passage of Lines	
	Occupy Assembly Area	

1. **Objective**. This FTX is designed to train the Infantry rifle company in conducting offensive operations. This exercise provides practice for the company commander, platoon leaders, squad leaders, and company headquarters personnel in planning, coordinating, and controlling combat operations.

2. **Interface**. This FTX is supported by STX C-II-1, C-II-2, and C-II-3 (Table 5-6). It is also supported by two warrior drills, *React to Contact* and *Break Contact*.

3. **Training**.

a. *Guidance*. FM 7-0 and FM 7-1 provide training guidance, tips, and enhancers for planning and executing STXs. The trainer should review the individual, leader, and collective tasks to be performed during the STX to determine which require initial or refresher training.

b. *General Situation*.

(1) The battalion is ordered to conduct offensive operations in support of the brigade mission. The company is ordered to infiltrate enemy lines and to destroy an insurgent operations base located in the CACTF. The company must be prepared to remain in the CACTF to conduct cordon and search missions.

(2) This exercise should be conducted under various conditions.

(3) When the company receives the battalion warning order the planning process for the attack mission begins. The exercise ends when the company completes a passage of lines, occupies an assembly area, and completes preparations to continue its mission. Figure 5-8 shows a graphic scenario of task performance for the FTX Attack. Table 5-7 provides a recommended sequence for performance of T&EOs and the time allotted for each portion of the STX, including AARs.

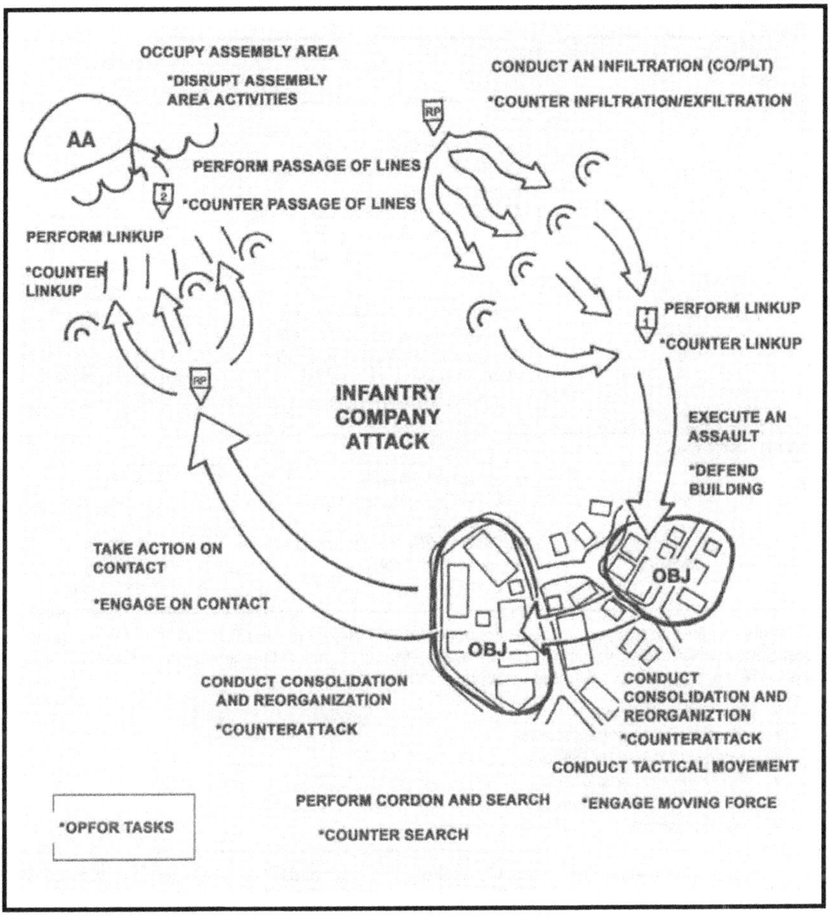

Figure 5-8. Attack FTX scenario.

Table 5-7. Sequence of events and times for FTX Attack.

Event	*Action*	*Estimated Time*
START STX C-II-1		
1	*Issue Company OPORD*	4 hours
2	*Conduct an Infiltration (Co/Plt)*	6 hours
3	*Perform Linkup*	3 hours
4	*Execute an Assault (MOUT) AAR*	3 hours
5	*Conduct Consolidation and Reorganization (Co/Plt)*	3 hours
ISSUE FRAGO, START STX C-II-2		
6	*Perform Cordon and Search*	6 hours
7	*Take Action on Contact / AAR*	3 hours
8	*Conduct an Infiltration (Co/Plt)*	6 hours
9	*Perform Linkup*	3 hours
10	*Perform Passage of Lines*	2 hours
11	*Occupy Assembly Area*	3 hours
START STX C-II-3		
12	*Disarm Belligerents*	3 hours
13	*Handle Captured Insurgents or Belligerents*	4 hours
14	*Apprehend/Detain Noncombatants*	4 hours
15	*Conduct Final AAR*	2 hours
	***Total Time**	**55 hours**

The time required to train an event varies based on the factors of METT-T and on the unit's training proficiency. Regardless, training is conducted to standards, not to time limitations. The following tasks are integrated and evaluated throughout the exercise:

- *Prepare for combat*
- *Maintain operations security*
- *Conduct tactical movement*
- *Perform consolidation*
- *Perform reorganization*
- *Perform stability and support operations*

* Time for intermediate AARs is excluded from this total; additional time is required if much of the exercise is conducted at night or during other limited visibility.

(4) The battalion commander issues an OPORD to the company commanders (Figure 5-9).

(Classification)

Copy no ___ of ___ copies
___ Battalion, ___ Infantry
Unit location or place of issue
Date-time group (with time zone)
Message reference number

OPERATION ORDER NO. _____
References: Map, series no. ___, sheet no. ___, edition ___
Time Zone Used Throughout the Order: ____

TASK ORGANIZATION:

___Company	___Company	___Company	Battalion Control
			Scout Platoon
			81-mm Mortar Platoon
			Antitank Platoon
			1/C/58th ENGR (DS)
			Battalion Trains

1. SITUATION.

a. **Enemy Forces**. Annex A (See intelligence overlay [prepared by brigade S-2].)

(1) *Disposition, Composition, and Strength*. The battalion is opposed by one company from the Dona Anna Militia Guards Battalion. This enemy company consists of 3 platoons with 35 to 40 Soldiers, equipped with M16s, AK-47s, and LMGs. The company headquarters is currently located in the town hall of CACTF. The company is at 90 percent strength and has indirect fire support assets available.

(2) *Capabilities*. The enemy possesses nonpersistent nerve agents and indirect fire assets capable of delivery. He also has some light armored vehicles in the area, but their location is undetermined. No enemy air threat exists. However, the enemy may be able to reinforce CACTF with an additional company, within two hours.

(3) *Most Probable Course of Action*. The enemy defends CACTF and continues using it as a base of operations and to consolidate weapons and material. His intent is to prevent the capture of CACTF and its support base, provided by its facilities and population.

(Classification)

Figure 5-9. Example battalion OPORD for FTX Attack.

(Classification)

b. **Friendly Forces.**

(1) (___) Brigade, (___) Division attacks at (date-time group) to seize the town of CACTF at (grid), to keep the enemy from using it as a base of operations. The intent is to secure the town and the road network in the area, to prevent the enemy from using it as a base of operations.

(2) (___) Battalion, (___) Infantry attacks to seize the road network at OBJ (____), vicinity (grid), at (date-time group) to prevent enemy forces from counterattacking along Route (____).

(3) (___) Battalion, (___) Infantry attacks at (date-time group), to seize the high ground at OBJ (____), vicinity (grid), and to establish defensive positions to block enemy routes in and out of CACTF.

2. **MISSION**. (___) Battalion, (___) Infantry attacks at (date-time group) to seize CACTF, vicinity (grid), NLT (date-time group), to prevent the enemy from using the town as an operations and logistics base.

3. **EXECUTION**.

a. **Concept of the Operation**. The battalion will accomplish its mission, which is to deny the enemy use of CACTF and its surrounding terrain and population, by seizing the town and controlling the key areas in the vicinity. We will conduct a feint in the west and then attack the decisive point within CACTF (OBJ [____]).

(1) *Maneuver*. (___) Company conducts a feint to deceive the enemy, then the main effort (*evaluated*) company seizes OBJ (____), to prevent the enemy from using CACTF as an operational and logistical base. (___) Company supports the main effort by seizing OBJ (____), in order to prevent the enemy from counterattacking the main effort. (___) Company seizes OBJ (____) to support the main effort's attack.

(2) *Fire Support*. Annex C. (See fire support.) Purpose of fires is to suppress the enemy and to obscure movement to CACTF. Priority of artillery fires is to (feinting) Company; on order, priority will shift to (supporting effort) Company. Priority of 81-mm mortar fires is to (*evaluated*) Company.

(3) *Engineering*. The purpose of the engineers is to breach obstacles during the attack, to help (*evaluated*) Company seize OBJ (____). Priority of support is to (*evaluated*) Company, (___) Company, and (___) Company in order. The battalion commander must authorize any booby traps, and point and hasty minefields.

(Classification)

Figure 5-9. Example battalion OPORD for FTX Attack (continued).

(Classification)

b. **Tasks to Maneuver Units.**

(1) *(Evaluated) Company.*

(a) Infiltrate using infiltration lane TANGO.

(b) Seize OBJ (____), vicinity (grid), NLT (date-time group).

(c) Be prepared to perform cordon and search of CACTF, after consolidation and reorganization.

(2) *(Supporting) Company.*

(a) Follow (main attack) Company and infiltrate on infiltration lane TANGO.

(b) Seize OBJ (____), vicinity (grid), NLT (date-time group).

(c) Coordinate the attack with *(evaluated)* Company, so that both units attack at the same time.

(d) Be prepared to perform cordon and search in the objective area after consolidation and reorganization.

(e) Be prepared to assume the mission of *(evaluated)* Company.

(3) *(Supporting) Company.*

(a) Infiltrate using infiltration lane ZEBRA.

(b) Conduct feint to OBJ (____), vicinity (grid), NLT (date-time group).

(c) Be prepared to establish defensive positions vicinity (grid) to prevent enemy movement to and from CACTF.

c. **Tasks to Subordinate Units.**

(1) *(Scout) Platoon.*

(a) During movement, determine location, strength, and disposition of enemy forces in CACTF.

(b) On order, screen the battalion flank from (grid) to (grid) to warn of the approach of enemy units from the southwest

(2) *Mortar Platoon.* Priority of mortar fire is to *(evaluated)* Company. Mortar platoon moves with *(evaluated)* Company, establishes a firing position vicinity (grid), and displaces on order.

(3) *Antitank Platoon.*

(a) Initially, over watch from (grid) to (grid) to secure the unit infiltration.

(b) On order, move vicinity (grid) to support *(evaluated)* Company's consolidation and reorganization.

(Classification)

Figure 5-9. Example battalion OPORD for FTX Attack (continued).

(Classification)

(4) *1/C/58 Engineers.*

(a) Move with (*evaluated*) Company to provide mobility support.

(b) Be prepared to reduce obstacles in CACTF during cordon and search mission.

d. **Coordinating Instructions.**

(1) Report when ready to execute an assault.

(2) Report any movement of enemy forces from the southwest.

(3) No rubbling is authorized, so minimize collateral damage to property.

(4) MOPP0 is in effect.

(5) Terrorist threat warning is red.

(6) Commanders brief back OPORD to battalion commander at battalion tactical CP at (date-time group).

(7) Rules of Engagement (ROE).

(a) Commanders take all steps necessary and appropriate for their units' protection.

(b) The minimum necessary force is used to control the situation.

(c) Leaders must take measures to minimize risk to civilians, without endangering the unit.

(d) Fire is returned directly to its source, not sprayed into a general area.

(e) Firing ceases when the threat is over.

(f) Anyone trying to surrender is allowed to do so.

(g) Civilians and property are treated with respect.

(h) WP can be used vicinity CACTF to aid in isolating the objectives. The requests for indirect fire, within the town, must be authorized by the battalion commander.

4. **SERVICE SUPPORT.** (See Annex D.)

5. **COMMAND AND SIGNAL.**

a. **Command.**

(1) The tactical CP moves with (*evaluated*) Company.

(2) The main CP is vicinity (grid) and displaces on order.

(3) The second in command remains in the battalion main CP.

(4) The rear CP is the alternate battalion main CP.

(Classification)

Figure 5-9. Example battalion OPORD for FTX Attack (continued).

(Classification)

b. **Signal.**

(1) Current SOI is in effect.

(2) Radio-listening silence is in effect for all elements except scouts, until prepared to attack.

ACKNOWLEDGE:
JONES
LTC
OFFICIAL:
SMITH
S-3
Annexes: A-Intelligence
B-Operations Overlay
C-Fire Support
D-Service Support

(Classification)

Figure 5-9. Example battalion OPORD for FTX Attack (continued).

(5) The battalion has destroyed the enemy operating in CACTF, and has seized the town. Two companies are defending against a limited counterattack. One company is in a blocking position. The (*evaluated*) company receives a FRAGO (Figure 5-10) to conduct a cordon and search operation. The purpose of the operation is to locate enemy personnel and equipment remaining in the CACTF.

Enemy personnel in squad-sized elements are reported by the civilian population to be hiding in the CACTF. Reports also indicate numerous weapons caches are being hidden by civilians collaborating with the enemy. Battalion has completed its mission and is preparing to conduct a cordon and search.

After completing this, the battalion exfiltrates to perform a passage of friendly lines. (Evaluated) Company moves to perform cordon and search vicinity OBJ (____), beginning (date-time group) to (date-time group), to locate and seize enemy personnel or equipment.

The (*evaluated*) company notifies the commander of the results of the search, and then notifies the commander when they (*evaluated* company) are ready to start movement. The company exfiltrates on order along Route Gold to perform linkup at (grid) and passage of lines at (grid). (See overlay for lanes.)

Figure 5-10. Example FRAGO for FTX Attack.

4. **Support Requirements**.

a. *Minimum Trainers/Evaluators*:. The company commander acts as the trainer-evaluator for the platoons. The battalion may direct the FTX internal evaluation along with a company evaluator and an FO evaluator. The platoon leaders and platoon sergeants are the primary trainers during the initial phases of the training. The company commander, the XO, the platoon leaders, and the platoon sergeants all act as trainers during the company phases of training. If the exercise is conducted by the battalion, either the battalion commander, XO, or S-3 acts as the trainer and evaluator, but additional battalion assets may be required to support the evaluation phase. The controllers or evaluators are positioned at critical locations throughout the exercise to observe the action. They must not interfere with the conduct of the operation, except to halt it when a safety violation occurs or to conduct an AAR.

b. *Vehicles/Communications*: Those organic to the company are used in the exercise. This includes one designated medical evacuation vehicle per platoon and at least one aidman on site.

c. *Opposing Force*:. The OPFOR ground force should be at least platoon-size. FTXs conducted by the company require internal resourcing; units should rotate to ensure total training. OPFOR tasks and standards are provided in each T&EO.

d. *Maneuver Area*: A training area 5 by 10 km which includes the CACTF is best for this exercise.

5. Evaluation.

Table 5-8 lists the T&EOs used to evaluate this STX.

Table 5-8. T&EOs used to evaluate STX Attack.

Task/Function	*Task Number*
Perform Cordon and Search	7-3-1105
Take Action on Contact	7-4-1107
Perform Passage of Lines	7-4-1125
Perform Linkup	7-4-1128
Conduct Tactical Movement	7-2-1342
Occupy Assembly Area	7-4-1136
Maintain Operations Security	7-4-1409
Prepare for Combat	7-4-1606
Conduct Consolidation and Reorganization (Co/Plt)	7-2-5027
Perform Cordon and Search	7-2-1105
Execute an Assault (MOUT)	7-2-1109
Perform Exfiltration	7-1-1146
Perform Infiltration	7-1-1147
Perform Reorganization	7-1-1608
Perform Consolidation	7-1-1609

MECHANIZED INFANTRY COMPANY OR TEAM FTX DEFEND

1. **Objective.** This FTX is designed to train the mechanized Infantry company or team in conducting defensive operations in urban terrain. It also provides platoons an opportunity to prepare for full-scale FTXs conducted and evaluated by the parent battalion. This exercise provides practice for the company commander, platoon leaders, squad leaders, and company headquarters personnel in planning, coordinating, and controlling combat operations.

2. **Interface.** This FTX is supported by the STX C-II-4, C-II-5, and C-II-6 tasks shown in Table 5-9. It is also supported by two warrior drills, *React to Contact* and *Break Contact.*

Table 5-9. Supporting tasks for company FTX.

STX C-II-4	*STX C-II-5*	*STX C-II-6*
Occupy Assembly Area	Perform Ambush	Operate a Checkpoint
Perform Tactical	Establish a Roadblock/Checkpoint	Interdict Smuggling Operations
Road March	Perform CBRN Operations	Defend Convoy
Perform Relief in Place	Conduct Tactical Movement	
Defend Built-Up Area/ Building	Perform Passage of Lines	
	Occupy Assembly Area	

3. **Training.**

a. *Guidance.* FM 7-0 and FM 7-1 provide training guidance, tips, and enhancers for planning and executing STXs. The trainer should review the individual, leader, and collective tasks to be performed during the STX to determine which require initial or refresher training.

b. *General Situation.*

(1) The battalion is ordered to conduct defensive operations in support of the brigade mission; the company is ordered to perform a relief operation and to prepare a defense in an urban area. The company remains in position for at least 36 hours before receiving a follow-on mission. Most civilians have been evacuated from the area, but a few remain. A terrorist threat exists as well as the possibility of enemy battalion-level offensive actions.

(2) This exercise should be conducted under various environmental conditions. The company also conducts operations under threat of CBRN attack.

c. *Special Situation.*

(1) The planning process for the defense mission begins when the company receives a warning order. The exercise ends when the company completes the passage of lines and occupies an assembly area. Figure 5-11 shows a graphic scenario of task performance for the FTX Defend.

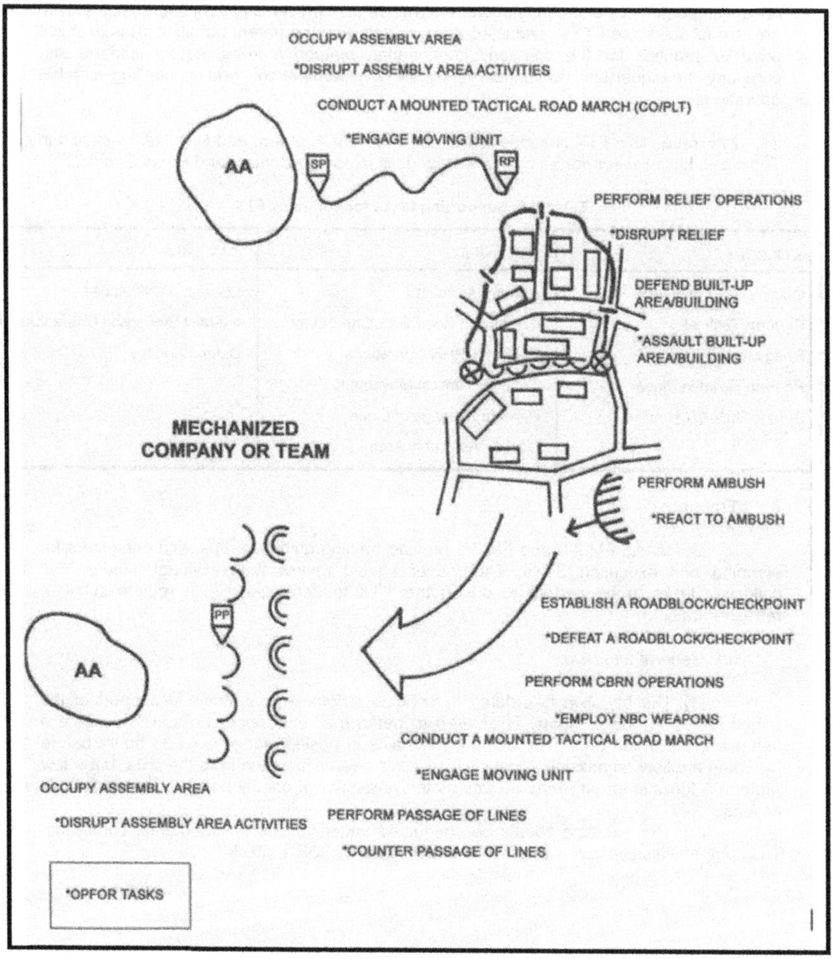

Figure 5-11. Graphic scenario of task performance for FTX Defend.

(2) Table 5-10 provides a recommended sequence for performance of T&EOs from DTMS and the time allotted for each portion of the STX, including AARs.

Table 5-10. Sequence of events and times for FTX Defend.

Event	*Action*	*Estimated Time*
START STX C-II-4		
1.	*Occupy Assembly Area*	2 hours
2.	*Issue Company OPORD*	4 hours
3.	*Perform Tactical Road March*	3 hours
4.	*Perform Relief Operations AAR*	6 hours
5.	*Defend MOUT/Building*	24 hours
ISSUE FRAGO, START STX C-II-5		
6.	*Perform Ambush*	6 hours
7.	*Establish a Roadblock/Checkpoint AAR*	6 hours
8.	*Perform CBRN Operations*	6 hours
9.	*Conduct Tactical Movement*	3 hours
10.	*Perform Passage of Lines*	2 hours
11.	*Occupy Assembly Area*	3 hours
START STX C-II-6		
12.	*Operate a Checkpoint*	2 hours
13.	*Interdict Smuggling Operations*	6 hours
14.	*Defend Convoy*	4 hours
15.	*Conduct Final AAR*	2 hours
The time required to train an event varies based on the factors of METT-T and on the unit's training proficiency. Regardless, training is conducted to standards, not to time limitations. The following tasks are integrated and evaluated throughout the exercise: ▪ *Prepare for combat* ▪ *Maintain operations security* ▪ *Conduct tactical movement* ▪ *Perform consolidation* ▪ *Perform reorganization* ▪ *Perform stability and support operations*		
* Time for intermediate AARs is not included in this total; additional time is required if much of the exercise is conducted at night or during other limited visibility.		

(3) The battalion commander issues the defense OPORD to the company (Figure 5-12).

```
                       ____________
                     (Classification)

                                            Copy no ___ of ___ copies
                                          ___ Battalion, ___ Infantry
                                      Unit location or place of issue
                                     Date-time group (with time zone)
                                             Message reference number
OPERATION ORDER NO. _____

Reference: Map, series no. ___, sheet no. ___, edition ___
Time Zone Used Throughout the Order: ____
TASK ORGANIZATION:
___Co ___Tm ___Tm (-) Battalion Control
 1/__/__/AR 2/__/__/AR-Scout Platoon
    120-mm Mortar Platoon
Team Tank E Co(-) 1/C/1-222 ADA (Stinger) (DS)
1/__/__/Inf  C/58 ENGR (DS)
2/__/__/Inf  1/1 (GSR) TM A/58 MI Bn
    ATK Force B (1/_/_/Inf)
    Battalion Trains
```

1. SITUATION.

a. **Enemy Forces.** Annex A (see intelligence overlay [prepared by brigade S-2]).

(1) *Composition, Disposition, Strength.* The company is opposed by the 1st Dragoons Regiment, 60th Guards Division. This regiment is at 85 percent strength and has received engineer and artillery reinforcements. The enemy is equipped with an assortment of former Soviet and Warsaw pact small arms, machine guns, and armored vehicles including T-62 tanks.

(2) *Capabilities.* The enemy has established defensive positions that follow a line from vicinity (grid) to vicinity (grid). Enemy armored reconnaissance units are operating in the more open terrain to the southwest of CACTF. The enemy has been observed moving forces in depth throughout the area in preparation for an offensive action.

(3) *Probable Course of Action.* The enemy's most probable course of action will be to begin offensive actions within the next 24 to 36 hours. Enemy encountered includes numerous reconnaissance elements trying to find the company's vulnerabilities and a main body trying to exploit any weakness.

____
(Classification)

Figure 5-12. Example battalion OPORD for mechanized company or team STX Defend.

(Classification)

b. **Friendly Forces.**

(1) (___) Brigade, (___) Division relieves (___) Brigade in place at (date-time group), to defend in sector from (grid) to (grid) to (grid) to (grid), in order to secure the corps counterattack axis of advance, (grid). The intent is to secure the road network around CACTF to prevent the enemy from impeding the movement of the (___) Brigade, (___) Division.

(2) TF (__) is the main effort in the defense and, NLT (date-time group), holds key terrain vicinity (grid) on our left, to prevent an enemy bypass of the brigade left flank.

(3) TF (__), on our right, defends along Route 360, vicinity (grid) to (grid) to (grid) to (grid), NLT (date-time group), to prevent an envelopment of the right flank. They conduct a supporting attack at (date-time group) to seize the high ground at OBJ OAK, vicinity (grid), to establish defensive positions that prevent the enemy armored reconnaissance units from interfering with the seizure of the pass at (grid) or the attack of the (___) Battalion, (___) Infantry, Mechanized.

2. MISSION. TF (____) defends NLT (date-time group), to retain the town of CACTF and its surrounding road network and allow brigade to prevent enemy interference with corps counterattack.

3. EXECUTION.

a. **Concept of the Operation**. Annex B (see operations overlay).

(1) Maneuver. Company (main effort) retains the main intersection (traffic loop) in CACTF, to prevent enemy movement through the sector. Team (*evaluated*) blocks movement west of 3d Street, to prevent the enemy from enveloping the main effort. The counterattack force is committed as needed. Team (__) blocks movement east of Oak Street, to canalize the enemy into EA (____). Team (__) defends from BP1, to prevent a coordinated attack against the main effort. E Company occupies BPs, to destroy enemy vehicles and protect the TF flanks.

(2) Fire Support. (See Annex C.) Fires are used to suppress the enemy and obscure our movement during the relief operation. Once positions are occupied, fires are used to suppress enemy overwatch positions, disrupt attempts to breach obstacles, and separate enemy Infantry from their supporting carriers. Priority of CAS and indirect fire is initially to Team (*evaluated*), during the relief operation, and then to Team (main effort). Teams (main effort) and (*evaluated*) each have two priority 155-mm targets. Each team may make one priority target an FPF.

(Classification)

Figure 5-12. Example battalion OPORD for mechanized company or team STX Defend (continued).

(Classification)

(3) Counter-Air Operations. Priority of protection is to the TF trains and main CP. Air defense weapons status is free; air defense warning status is yellow.

(4) Intelligence. Priority of collection is to identify and locate the C2 elements of the MRR and MRB, enemy tank battalions, and the enemy assault detachments' actual axes of advance.

(5) Engineering. The purpose of obstacles is to turn the enemy into restrictive terrain and engagement areas. Priority of support is to Team (main effort), Team (*evaluated*), Team (__), and Team (__), in that order.

b. **Tasks to Maneuver Units.**

(1) Team (Main Effort).

(a) Sector of defense includes the east side of Front Street, the south side of 1st Street, the west side of 3d Avenue area, and the north side of 3d Street.

(b) Team (main effort) relieves Team (__), TF (__).

(2) Team (Evaluated).

(a) Sector includes the east side of 3d Avenue, the south side of 1st Street, the west side of 6th Avenue, and the north side of 3d Street.

(b) Team (*evaluated*) relieves Team (__), TF (__).

(3) Team (__).

(a) Sector includes the east side of Oak Street, the south side of 1st Street, the east side of Front Street, and the north side of 3d Street.

(b) Team (__) relieves Team (__), TF (__).

(c) Team (__) coordinates directly with right flank team of TF (__), to ensure mutually supporting fires.

(4) Team (__)

(a) Occupy BP 1 vicinity (grid)

(b) Be prepared to occupy BP 2 (grid) to engage enemy in EA (____).

(5) Company E.

(a) Occupy BP 2-1 (grid) and move on order to BP 2-2 (grid).

(b) Provide antiarmor fires to destroy enemy tanks in EA (____).

(6) Counterattack B (Reserve).

(a) Occupy AA Black, vicinity (grid).

(b) Reconnoiter routes into TF (main effort) sector.

(Classification)

Figure 5-12. Example battalion OPORD for mechanized company or team STX Defend (continued).

(Classification)

(3) Counter-Air Operations. Priority of protection is to the TF trains and main CP. Air defense weapons status is free; air defense warning status is yellow.

(4) Intelligence. Priority of collection is to identify and locate the C2 elements of the MRR and MRB, enemy tank battalions, and the enemy assault detachments' actual axes of advance.

(5) Engineering. The purpose of obstacles is to turn the enemy into restrictive terrain and engagement areas. Priority of support is to Team (main effort), Team (*evaluated*), Team (__), and Team (__), in that order.

b. **Tasks to Maneuver Units.**

(1) Team (Main Effort).

(a) Sector of defense includes the east side of Front Street, the south side of 1st Street, the west side of 3d Avenue area, and the north side of 3d Street.

(b) Team (main effort) relieves Team (__), TF (__).

(2) Team (Evaluated).

(a) Sector includes the east side of 3d Avenue, the south side of 1st Street, the west side of 6th Avenue, and the north side of 3d Street.

(b) Team (*evaluated*) relieves Team (__), TF (__).

(3) Team (__).

(a) Sector includes the east side of Oak Street, the south side of 1st Street, the east side of Front Street, and the north side of 3d Street.

(b) Team (__) relieves Team (__), TF (__).

(c) Team (__) coordinates directly with right flank team of TF (__), to ensure mutually supporting fires.

(4) Team (__)

(a) Occupy BP 1 vicinity (grid)

(b) Be prepared to occupy BP 2 (grid) to engage enemy in EA (____).

(5) Company E.

(a) Occupy BP 2-1 (grid) and move on order to BP 2-2 (grid).

(b) Provide antiarmor fires to destroy enemy tanks in EA (____).

(6) Counterattack B (Reserve).

(a) Occupy AA Black, vicinity (grid).

(b) Reconnoiter routes into TF (main effort) sector.

(Classification)

Figure 5-12. Example battalion OPORD for mechanized company or team STX Defend (continued).

(Classification)

(c) Priority for planning:
- Battle position to reinforce main effort.
- Battle position to counterattack into Team (main effort) sector.
- Battle position to counterattack into Team (*evaluated*) sector.

(1) Scout Platoon.
(a) Determine location, strength, and disposition of enemy forces.
(b) Perform a screen of the TF's right flank from (grid) to (grid) beginning (date time group) to prevent enemy infiltration of CACTF.

c. **Tasks to Subordinate Units.**
(1) Mortar Platoon.
(a) Priority of fire to TF (main effort). Move with TF (main effort) and establish firing position vicinity (grid).
(b) Displace on order to (grid).
(2) 1/C/1 222 ADA. Coordinate locations with S-3 NLT (date time group).
(3) C/58 Engineers.
(a) Coordinate the engineer effort with the team commanders NLT (date time group).
(b) Provide an obstacle clearing team to Counterattack Force B, NLT (date time group).
(4) 1/1 (GSR) TM A/58 MI. Coordinate location of assets with the task force S-3 NLT (date time group).

d. **Coordinating Instructions.**
(1) Selective rubbling is authorized, but destruction of more than one building requires TF commander approval.
(2) Destruction of subterranean approaches requires TF commander approval.
(3) Commanders brief back OPORD to TF commander at the TF tactical CP, vicinity (grid), at (date time group).
(4) MOPP1 is in effect as of (date time group).
(5) Decontamination site established vicinity Miller Pond (grid).

(Classification)

Figure 5-12. Example battalion OPORD for mechanized company or team STX Defend (continued).

(Classification)

4. SERVICE SUPPORT. (See Annex D.)

a. **General.**
(1) Combat trains located vicinity (grid).
(2) Field trains located in the BSA, vicinity (grid).

b. **Materiel and Services.**
(1) Supply.
(a) Class I. Ration cycle C-A-C. Team establishes MRE cache sufficient for 48 hours in sector.
(b) Class III. Resupply available on request. Class III located in combat trains.
(c) Class IV. Materials are allocated as shown in Annex D.
(d) Class V. Each team establishes ammunition caches in sector, but should also have two basic loads of ammunition on hand.
(2) Transportation. The main streets are capable of two way heavy vehicle traffic (class 60). Secondary streets may be one way and of lighter construction (class 50).
(3) Maintenance.
(a) Priority by unit: Team (main effort), Team (*evaluated*), Team (__), Company E, Team (__).
(b) Priority by vehicle: M1A1, M2/M3, M113.

c. **Medical Evacuation and Hospitalization.**
(1) TF aid station located vicinity (grid).
(2) Civilian casualties should be transported by nonmilitary means to clinic, located vicinity (grid).

d. Personnel.
(1) No replacements for the next 96 hours.
(2) EPW collection point located vicinity (grid).
(3) Mortuary affairs point located in BSA, vicinity (grid).

e. **Civil-Military Cooperation.**
(1) Curfew in effect 1800 to 0800 daily.
(2) Majority of civilians have been evacuated. Those remaining have been notified of the risk involved.

(Classification)

Figure 5-12. Example battalion OPORD for mechanized company or team STX Defend (continued).

(Classification)

5. **COMMAND AND SIGNAL.**

a. **Command.**
(1) Brigade main CP located vicinity (grid).
(2) TF tactical CP located vicinity (grid).
(3) TF main CP located vicinity (grid).
(4) TF commander located at tactical CP.
(5) TF XO located at main CP.

b. **Signal.**
(1) SOI index 1 9 in effect.
(2) Primary means of communication is wire, FM, messenger in that order.
(3) Building and obstacle marking IAW TF SOP.

ACKNOWLEDGE:

WILLIAMS
LTC

OFFICIAL:

SMITH
S-3

Annexes: A–Intelligence
B–Operations Overlay
C–Fire Support
D–Service Support

(Classification)

Figure 5-12. Example battalion OPORD for mechanized company or team STX Defend (continued).

(3) The TF has retained the north half of the CACTF after the enemy attack. The enemy was defeated but has established squad-sized to platoon-sized strongpoints in the southern half of the CACTF. The enemy is expected to receive reinforcements within the next 12 to 24 hours. Theprevent enemy reinforcements from moving to the strongpoints to strengthen and resupply them, TF headquarters orders the team to conduct platoon-sized ambushes. They also order the team to prepare to establish a roadblock, to control refugee flow through the area of operations, before the team moves out of the area. Headquarters issues a FRAGO (Figure 5-13).

```
Numerous enemy strongpoints are located to your front, vicinity (grid), (grid), and (grid). We expect these strongpoints to receive reinforcements within the next 12 to 24 hours.

Your mission is to establish platoon ambushes on the likely avenues of approach to the strongpoints, NLT (date-time group), in order to prevent reinforcements or resupply to these strongpoints.

Notify the TF commander when you are prepared to move.

Coordinate with TF scouts for guides. On order, after return from ambush missions, establish roadblocks at (grid) and (grid). Be prepared to move within the next 36 to 48 hours to conduct further follow-on missions.
```

Figure 5-13. Example FRAGO for FTX Defend.

4. **Support Requirements.**

a. *Minimum Trainers/Evaluators*: This exercise should be conducted for practice with the company commander as the trainer-evaluator for his platoons. The battalion may direct the FTX (internal evaluation) using a company evaluator and an FO evaluator. The platoon leaders and platoon sergeants are the primary trainers during the initial phases of the training. The company commander, the XO, the platoon leaders, and the platoon sergeants act as trainers during the company phases of training. If the exercise is conducted by the battalion, either the battalion commander, the XO, or the S-3 acts as the trainer and evaluator, but may require additional assets from the battalion to support the evaluation phase. The controllers or evaluators must be positioned at critical locations throughout the exercise, to observe the actions. They must not interfere with the conduct of the operation, except to halt it when a safety violation occurs or to conduct an AAR.

b. *Vehicles/Communications*: Those organic to the company are used.

c. *Opposing Force*:. The OPFOR ground force should be at least platoon-size. FTXs conducted by the company require internal resourcing, and units should be rotated to ensure total training. T&EOs provide training tasks and standards for the OPFOR.

d. *Maneuver Area*: A training area at least 5 by 15 km is best for this exercise.

e. *Consolidated Support Requirements*: This exercise requires the items shown in Table 5-11.

f. *Barrier materials.* Sufficient Class IV must be made available to fortify positions, IAW FM 3-06.11.

Table 5-11. Ammunition and pyrotechnics support requirements for a six-day FTX.

Ammunition	*DODAC*	*Per Plt/Co/Bn FTX*
5.56-mm blank	1305A080	120 rounds per rifle
5.56-mm SESAMS		120 rounds per rifle
5.56-mm blank	1305A075	200 rounds per M249 LMG
5.56-mm SESAMS		200 rounds per M249 LMG
7.62mm	1305A111	200 rounds per BFV/M1
		400 rounds per M60/240B
ATWESS cartridge	1370L367	6 (Viper) (for LAW) 18 per Dragon 14 per TOW system 9 per OH-58D/AH-64
Pyrotechnic device	1370L602	6 per M1 2 per target
Hand grenade fuze (M228)	1330G878	150 per battalion
Stun grenade M84	1330GG09	150 per battalion
Smoke grenade, HC		150 per battalion
Smoke pot		6 per battalion

5. **Evaluation**.

Table 5-12 lists the DTMS T&EOs used to evaluate this STX.

Table 5-12. T&EOs used to evaluate STX Attack.

Task/Function	*Task Number*
Defend MOUT/Building	7-3/4-1118
Conduct a Dismounted Tactical Road March (Co/Plt)	7-2-1189
Conduct a Mounted Tactical Road March (Co/Plt)	7-2-1198
Perform Relief Operations	7-3/4-1124
Perform Passage of Lines	7-3/4-1125
Conduct Tactical Movement	7-2-1342
Occupy Assembly Area	7-3/4-1136
Establish a Roadblock/Checkpoint	7-3/4-1401
Perform CBRN Operations	7-3/4-1406
Maintain Operations Security	7-3/4-1409
Prepare for Combat	7-3/4-1606
Conduct Consolidation and Reorganization	7-2-5027
Perform Ambush	7-2-1106

TARGETRY

5-10. Table 5-13 lists the combination of precision and nonprecision targets to support the diversity of scenarios required to support combined arms and branch specific lane training. Precision targets, less than 50 meters, are used in and around buildings within the CACTF and reinforce precision marksmanship techniques described in FM 3-22.9 (Table 5-13).

Table 5-13. CACTF target requirements.

Event	*Number Of Lanes*	*Human Targets*		*Vehicle Targets*	*Totals*	
					HT	*VT*
Company/Platoon STX Offense	3/9	Precision	Nonprecision	9	45	9
		15	30			

CONSIDERATIONS FOR TRAINING WITH ARMOR

5-11. Built-up areas consist mainly of man-made features such as buildings, streets, and subterranean systems. These features of urban terrain create a variety of tactical problems and possibilities. In order to ensure that the tank platoon can operate effectively in the urban environment, the platoon observation and direct-fire plans must address the ground-level or surface fight (in streets and on the ground floor of buildings), the aboveground fight (intrasurface and supersurface), and the subterranean fight (Figure 5-14). Infantry can assist the tank platoon by—

- Locating targets for tanks to engage.
- Destroying antitank weapons.
- Assaulting enemy positions and clearing buildings with tank support.
- Protecting tanks from antitank fires.

Figure 5-14. Urban battlespace.

Transporting Infantry

5-12. At times, the tank platoon must transport Infantrymen (Figure 5-15). They only do so when not expecting contact. If the platoon is moving as part of a larger force, and if it is tasked to provide security for the move, then the lead section or element does not carry Infantry.

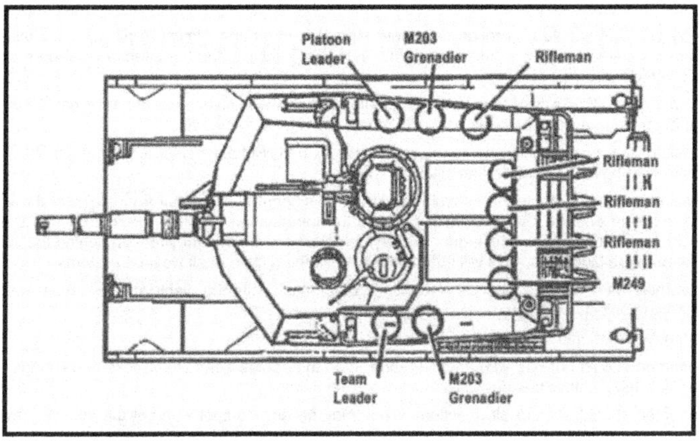

Figure 5-15. Example positions for Infantry riding on a tank.

5-13. Infantry and Armor leaders must observe the procedures, precautions, and considerations in Figure 5-16 when Infantrymen ride on tanks:

- Infantry should thoroughly practice mounting and dismounting procedures and actions on contact.
- Infantry must always alert the TC before mounting or dismounting. They must follow the commands of the TC.
- Infantry platoons should be broken down into squad-sized groups, similar to air assault chalks, with the Infantry platoon leader on the Armor platoon leader's vehicle and the Infantry PSG on the Armor PSG's vehicle.
- Platoon leaders, PSGs, and team leaders should position themselves near the TC's hatch, using the external phone (if available) to talk to the TC and relay signals to the unit.
- If possible, the lead vehicle should not carry Infantrymen. Riders restrict turret movement and are more likely to be injured or killed on initial contact.
- Whenever possible, Infantrymen should mount and dismount over the left front slope of the vehicle. This procedure ensures that the driver can see the Infantrymen and that the Infantrymen DO NOT pass in front of the coax machine gun. Infantrymen must ensure that they remain behind the vehicle's smoke grenade launchers. This will automatically keep them clear of all weapon systems.
- Infantrymen must always have three points of contact with the vehicle; they must watch for low-hanging objects such as tree branches.
- All Infantrymen should wear hearing protection.
- Infantrymen should not ride with anything more than their battle gear. Rucksacks and B-bags should be transported by other means.
- Infantrymen should scan in all directions. They may be able to spot a target the vehicle crew does not see.

Figure 5-16. Precautions for Infantrymen riding on a tank.

5-14. Passengers should be prepared to take the actions on contact shown in Figures 5-17 and Figure 5-18.

- Wait for the vehicle to stop.
- At the TC's command, dismount immediately (one fire team on each side). ***Do not*** move forward of the turret.
- Move at least 5 meters to the sides of the vehicle.
  Do not move behind or forward of the vehicle.
- ***Do not*** move in front of vehicles unless ordered to do so. Main gun discharge overpressure can inflict severe injury or death to forward dismounts.
- ***Do not*** dismount a vehicle unless ordered or given permission to do so.
- ***Do not*** dangle arms or legs, equipment, or anything else off the side of a vehicle; they could get caught in the tracks, causing death, injury, or damage to the equipment or vehicle.
- ***Do not*** place too many riders on the vehicle.
- ***Do not*** fall asleep when riding. The warm engine may induce drowsiness; a fall could be fatal.
- ***Do not*** smoke when mounted on a vehicle.
- ***Do not*** stand near a moving or turning vehicle at any time. Tanks have a deceptively short turning radius.

Figure 5-17. Actions on contact for Infantrymen riding on a tank.

DANGER

1. The overpressure from the tank 120-mm cannon can kill a dismounted Infantryman within a 90-degree arc extending from the muzzle of the gun tube out to 200 meters.
2. From 200 to 1,000 meters along the line of fire, on a frontage of about 400 meters dismounted Infantry must be aware of the danger from discarding sabot petals, which can kill or seriously injure Soldiers.

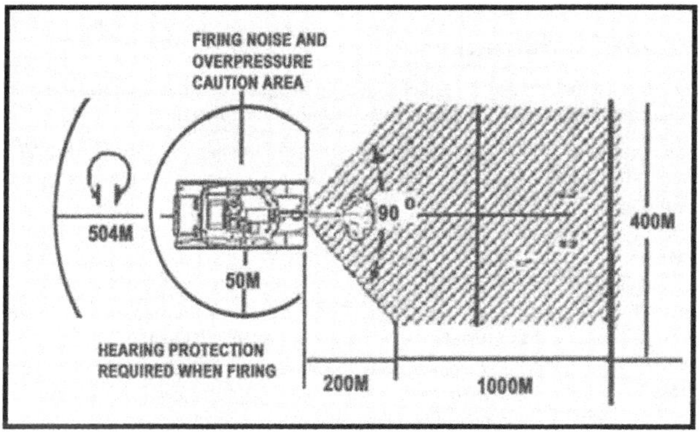

Figure 5-18. Danger areas around a tank when firing the 120-mm main gun.

ADDITIONAL CONSIDERATIONS

5-15. Additional considerations and preparations for transporting Infantrymen include the following:

- Armor--
  - Uses main gun fire to reduce obstacles or entrenched positions for the Infantry.
  - Takes directions from the Infantry ground commander (platoon leader/PSG/squad leader) to support their fire and maneuver.
  - Provides reconnaissance by fire for the Infantry.
  - Should know and understand how the Infantry clears buildings, how they mark cleared buildings, casualty evacuation plan, signal methods, engagement criteria for tank main gun, front line trace reporting, ground communication from the tank with the dismounted Infantry.
  - Uses its night vision capability to augment and supplement the Infantry's night vision capability.
- Infantry--.
  - Provides real time information for the tank crewmen to help them overcome tank noise and the lack of ground situational understanding.
  - Provides reconnaissance and fire direction of enemy positions for main gun attack.

- Considerations for dismounted tank security include--
  - -- Tank crewman should rehearse the mounting and dismounting of Infantry from their vehicle, briefing the Infantrymen on safety procedures for the vehicle and weapon systems.
  - -- Tank commanders need to rehearse communicating with dismounted Soldiers by way of TA-1 and DR-8 in the bustle rack.
- Vehicle preparation for combat in urban terrain should cover--
  - -- Keep at least one ballistic shield to the "Dog House" closed. Most engagements will be under boresight range, and the battlesight technique will suffice.
  - -- Place sandbags around antenna connections and electrical wiring on the turret top.
  - -- Place extra coax ammunition inside the turret.
  - -- Remove any highly flammable products from the outside of the vehicle and from the sponson boxes.

VEHICLES, WEAPONS, AND MUNITIONS

5-16. Numerous factors related to vehicles and their organic weapons and munitions affect the tank platoon's urban operation planning and execution, including the following:

- The preferred main gun rounds in the urban environment are HEAT, MPAT (ground mode), and MPAT-OR. These all perform much better than sabot rounds against bunkers and buildings.
- HEAT ammunition will open a larger hole in reinforced concrete or masonry structures than MPAT or MPAT-OR. Both MPAT and MPAT-OR, however, offer greater incapacitation capability inside the structure.
- HEAT ammunition arms approximately 60 feet from the gun muzzle. It loses most of its effectiveness against urban targets at ranges of less than 60 feet.
- MPAT and MPAT-OR rounds arm approximately 100 feet from the muzzle of the gun. Because of the shape and metal components of the projectiles, however, this ammunition remains effective at ranges of less than 100 feet.
- Sabot petals, including those on MPAT and MPAT-OR, endanger accompanying Infantry elements. They create a hazard area extending 70 meters on either side of the gun-target line out to a range of 1 kilometer.
- The tank's main gun can depress only to -10 degrees and can elevate only to +20 degrees. This creates considerable dead space for the crew at the close ranges that are typical in the urban environment.
- The external M2 HB machine gun can deliver a heavy volume of suppressive fire and penetrate light construction, buildings and most barricades. The M2 HB MG can elevate to +36 degrees; however, the TC must be unbuttoned to fire the M2 on the M1A2 or M1A2 SEP.
- The M240 coax machine gun can effectively deliver suppressive fires against enemy personnel and against enemy positions that are behind light cover.
- The loader's M240 machine gun can effectively deliver suppressive fire against enemy personnel and against enemy positions that are behind light cover; however, the loader must be unbuttoned to operate it. This weapon may be dismounted and used in a ground role if units are equipped with the M240 dismount kit.

- When buttoned up, the tank crew has limited visibility to the sides and rear and no visibility to the top. Figures 5-19 and 5-20 show the dead spaces associated with tank operations in an urban environment.
- FM 3-20.12 explains special uses for tank-mounted machine guns in the urban environment.

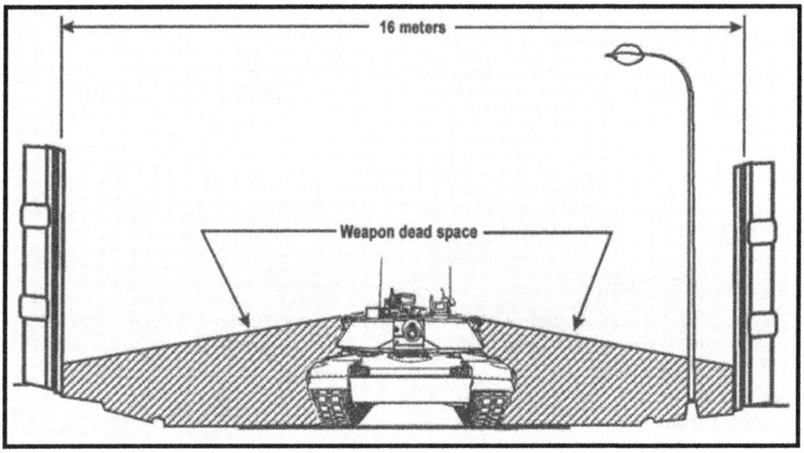

Figure 5-19. Tank weapon dead space at street level.

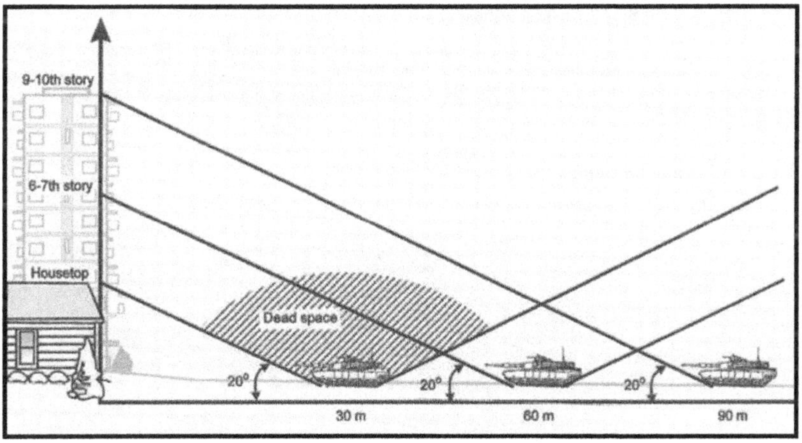

Figure 5-20. Tank main gun and coax dead space above street level.

Tank Platoon Command and Control in Urban Operations

5-17. Communications problems, fire control, proximity and visibility, personnel factors, ROE, ROI, civilians, and slow-pace urban operations all affect the tank platoon's urban operations planning and execution:

Communications Problems

5-18. The low-level task organization that may take place during urban operations requires elements to establish additional communications links, which can be disrupted by buildings and other urban terrain features.

Fire Control

5-19. Extensive direct fire planning and restrictive fire control measures are an absolute requirement in urban operations.

Proximity and Visibility

5-20. Friendly elements often must operate in confined and restrictive areas during urban operations, and they may not be able to see other nearby friendly forces. These factors significantly increase the danger of fratricide.

Personnel Factors

5-21. Urban operations impose significant, and often extreme, physical and psychological demands on Soldiers and leaders.

ROE, ROI, and Civilian Considerations

5-22. The ROE or ROI may restrict the use of certain weapon systems and TTP. As an integral part of urban operations, noncombatants create special operational problems. Thedeal with these concerns, units operating in urban terrain must know how to effectively employ linguists and counterintelligence and civil affairs teams.

Slow Pace Urban Operations

5-23. This operation usually prevents the tank platoon from taking full advantage of the speed and mobility of its tanks.

Appendix A

Training Tips

The effectiveness of urban operations training depends on the imagination and thoroughness applied to a training plan. All trainers, from training managers to junior leaders, should prepare their units to fulfill their urban operations mission. They can employ procedures that are based on training doctrine and proven techniques. This appendix discusses several techniques for conducting urban operations training. Trainers make the choice based on the training needs and available resources.

CONCURRENT TRAINING

A-1. Concurrent training allows groups of Soldiers to train on different tasks at the same time. This training requires a plan for rotating Soldiers between the stations that teach different tasks. This procedure is referred to as round-robin training. Concurrent training is an efficient training technique since it makes the best use of training time and facilities. Concurrent training in the urban operations training facilities works best with performance-oriented training. It also requires detailed planning, control, coordination, and assistant trainers. Each station should conduct training that is relevant to, and supports the training the unit is conducting.

URBAN ASSAULT COURSE AND BREACH FACILITY

A-2. These facilities are well suited for conducting concurrent training. After mastering the tasks at one station, a group of Soldiers rotates to the next station in the sequence. Theavoid bottlenecks at any of the stations, trainers plan instructional blocks that last about the same length of time. Concurrent training is often conducted with dedicated instructors at each station. The *advantage* of this technique is that it ensures uniformity of performance standards and saves on trainer preparation time. The *disadvantage* is that the Soldier is not always trained by his leader. Therefore, when feasible, small-unit leaders should rotate with their Soldiers from station-to-station so they can take part in the training. This technique demands prior planning, preparation, and instructors that can teach.

URBAN ASSAULT COURSE AND COMBINED ARMS COLLECTIVE TRAINING FACILITY

A-3. These facilities are well suited for conducting concurrent exercises. Concurrent exercises are variants of concurrent training. In a concurrent exercise, elements of the same unit receive different missions. These missions correspond to each element's training objective for that exercise. This technique can be used during FTXs, STXs, or CFXs. For example, one unit attacks; a second unit supports by fire; a third unit acts as a reserve. Also, elements of the same unit can challenge each other. For example, if a trainer has been training his unit in urban offense and defense operations, he can create a scenario in which one force is required to attack the other. This technique allows the unit to train with an OPFOR without an extra investment of resources. Leaders must remember to carefully plan the force ratios between the OPFOR and counterattacking forces.

DEMONSTRATIONS

A-4. Trainers must demonstrate the correct way to perform the tasks that the Soldiers will be required to perform. A demonstration is more effective when the trainer uses hand-on techniques. A live demonstration is an effective teaching method since it holds Soldiers' interest and they are able to learn with their eyes. The UAC and the Breach Facility are ideal facilities for teaching individual and collective tasks using demonstrations. The trainer must ensure that each Soldier. Performs the task being taught to guarantee that the soldier has learned. Key points should be explained to the satisfaction of each Soldier before progressing to the next step. The trainers can use video cameras to film the action occurring inside the buildings to show the Soldiers the correct way to perform a task or use the same action to explain a better technique to accomplish the same task. This technique is very effective because the Soldiers can watch the demonstrations on a remote monitor, and trainers can replay the videotape as often as required to reinforce his training objective.

EXERCISES

A-5. Training exercises are used to develop tactical SOPs, teamwork among the participants, and Soldier confidence in his leaders and equipment. Table A-1 matches training exercise types to appropriate training facilities.

Table A-1. Training exercise matrix.

EXERCISE	*CACTF*	*SHOOT HOUSE*	*SIMULATIONS*	*LOCAL TOWN*
MAPEX	X		X	X
TEWT	X			X
CPX	X		X	X
CFX	X			X
BSX			X	
LFX		X		
STX	X	X		X
FTX	X			X

MAP EXERCISE

A-6. A MAPEX allows participants to plan and execute urban operations using maps, overlays, sand tables, photographs, or models. MAPEXs develop teamwork and skills in terrain analysis, planning, coordination, order writing, and employment of urban operation tactics and are more suited for battle staff and leaders.

- Using a model of the CACTF is an excellent training aid for instructing Soldiers and leaders. Gather leaders around the map or model and have them devise a tactical plan. Use figures or markers to represent friendly units and the OPFOR. Create tactical situations that challenge participants to react as they would in real combat.
- Incorporate direct fire, casualties, CBRN, logistics, and civilians into the exercise. Ensure the ROE for each exercise is clear to participants.
- Have plenty of detailed maps available for use in the MAPEX and in the planning and execution of subsequent CFX operations.

TACTICAL EXERCISE WITHOUT TROOPS

A-7. A TEWT focuses on leader and staff tasks. The senior trainer selects the terrain for the urban TEWT. Either the CACTF or a real village, town, or city is appropriate. The procedure for conducting a TEWT is as follows:

- Create a scenario and enemy situation, and assign missions.
- Allow the leaders to devise their reconnaissance plans and implement them on the terrain.
- Require the leader to brief subordinates while walking the terrain.
- Ensure their plans include the following tasks:
  -- Analyze the terrain.
  -- Employ units according to the terrain analysis.
  -- Emplace weapons.
  -- Devise a unit plan.
- Conduct an AAR.

COMMAND POST EXERCISE AND BATTLE SIMULATION EXERCISE

A-8. CPXs and BSXs allow leaders and battle staffs to perform their tasks in a unit CP, where they receive information as reports requiring certain reactions. This type of training can be achieved through the constructive simulations training portion of the urban operations training strategy using JANUS or JCATS. A company CP is usually associated with a battalion level or higher CPX, but a company can conduct its own CPX. The headquarters section can set up a CP in a CACTF building, displace to another building, and repeat the exercise until they meet the required standards. The section can practice hiding wire on pavement, stringing it between buildings, and running it through sewers. The CP can practice receiving reports, posting situation maps, and sending reports forward to higher headquarters. The CPX/BSXs are most effective in achieving the following general training objectives:

-- Preparing plans.
-- Issuing orders.
-- Building teams.
-- Reconnoitering, selecting, and tactically occupying CP locations.
-- Establishing and using communications.
-- Preparing and sending reports.
-- Coordinating between subordinate elements.

COMMAND FIELD EXERCISE

A-9. The CFX is used to train commanders, staffs, and junior leaders. The CACTF or a local town can accommodate this type of exercise. However, the unit commander must go through the correct procedures when conducting training off federal property. The CFX employs all of a unit's command and control assets, with a number of subordinate elements reduced according to a set scale. For example, one Soldier might represent an entire squad, one vehicle a platoon, or one gun an entire battery.

Preliminary Training

A-10. Preliminary training, such as CPXs, TEWTs, MAPEXs, and BSXs, should precede the CFX. Units conduct operations, and controllers resolve battle outcomes. The exercise ends with an AAR. Fewer controllers are needed in a CFX than in an FTX. However, if the number of unit functions increases, the number of controllers must increase also.

Planning Procedures

A-11. Combat in urban areas entails centralized planning with decentralized execution.

Centralized

A-12. Centralized planning can be practiced in a CFX; however, it can be practiced more efficiently in a CPX, TEWT, MAPEX, or BSX.

Decentralized

A-13. Decentralized execution means that platoon, squad, and team leaders lead the battle in an urban area. However, because element participation is scaled down in a CFX, trainers usually find STXs, FTXs, and TEWTs more useful.

Live-Fire Exercise

A-14. Chapter 3 and local range regulations provide specific guidance for conducting LFXs at the shoot house. Live-fire training allows Soldiers and leaders to maneuver while firing service ammunition or SRTA. An LFX is most productive when preceded by non-live-fire practice (dry fire or dry run). However, dry fire should *always* precede live fire, with either TES or SESAM. An AAR should follow each repetition, and units should only proceed to the next level of difficulty after mastering the current step. The training in an LFX occurs in the following sequence, which can be shortened if time and ammunition dictate.

1. Dry fire to practice unit drills and individual tasks.
2. Conduct an AAR.
3. Practice drills, tasks a second time with limited use of ammunition to show the complex coordination of fire and maneuver required.
4. Conduct an AAR.
5. Practice exercise a third time with full use of ammunition to enhance realism and build confidence.
6. Conduct an AAR.

Situational Training Exercise

A-15. STXs are mission-oriented, limited exercises used to train a unit to perform a single collective task or a series of related tasks and drills. With their distinct start and stop points, STXs represent a segment of battle. An STX is more flexible and complex than a single T&EO or drill. In fact, an STX typically includes drills, leader tasks, and separate individual tasks, and involves a unit's "slice" of external combat assets. An STX often precedes an FTX or CFX, and can be conducted in the CACTF by creating a brief situation and mission. For example, the MP company commanders might want to teach his Soldiers about straggler control. They can string the tasks together into small scenarios such as STXs, describe a tactical situation that produces stragglers from nearby units, use other Soldiers to represent the stragglers, and train Soldiers to process stragglers at straggler collection and transport points.

Field Training Exercise

A-16. FTXs allow units to execute collective tasks under conditions that closely resemble actual urban terrain. FTXs are suitable for the CACTF.

- An FTX requires significant time, resources, and planning. The time is well spent if units master their collective tasks. An FTX is an excellent way to conduct the *run* stage of collective training. Additionally, OPFOR and MILES greatly enhance the training value of the FTX.
- A unit's urban operations FTX training plan addresses ROE, maneuver restrictions, and the controller plan. Soldiers must know the exercise's training objectives before they begin, so they can focus on the correct tasks. Every participant, regardless of position or rank, must fully understand all safety requirements.

- To make the most of their training time in the field, leaders must plan in great detail. The concept of multiechelon training demands that they consider what training can occur at each level in each event.

EVALUATORS

A-17. Soldiers from the training unit or from another unit can evaluate urban operations training, especially during the *run* stage of training in an STX, FTX, or CFX. A dedicated evaluator provides objective and thorough comments on the unit's performance throughout all stages of the exercise. Evaluators monitor the critical point of the action but must avoid interfering. They must not disrupt unit performance or undermine realism. They must be trained and given standardized evaluation guides if their feedback is to be meaningful; MTPs and T&EOs are a good start. Evaluators facilitate AARs alone or along with the unit leader.

TACTICAL ENGAGEMENT SYSTEM

A-18. The TES realistically replicates the effects of actual fire in training. It uses laser adapters on the weapons and laser-sensitive equipment on the Soldiers and vehicles. A TES weapon emits a laser beam when fired. Accurate fire strikes the laser-sensitive equipment and scores a hit. The TES requires Soldiers to aim accurately and to use cover and concealment effectively. When used with an OPFOR, TES offers realistic training. It is an excellent training aid for an FTX or STX. It is also effective at the shoot house as a substitute for live fire. The two major shortcomings of TES are that it allows Soldiers to seek cover behind structures that normally would not protect them, and that it does not replicate suppression.

NAVIGATION

A-19. Navigational techniques for urban areas differ from those needed on other terrain.

- A typical 1:50,000-scale military map provides insufficient detail for accurate navigation. Therefore, Soldiers must learn to navigate using commercial maps, sketches, and aerial photographs. Leaders must learn how to plan operations using these items.
- Underground navigation is a special challenge. Soldiers must know subway plans and be able to identify tunnels, elevated tracks, stairways, and other critical features. They must learn how to measure distance, remain oriented, and mark their routes.
- Destruction often erases urban features that could aid in navigation. Soldiers must learn to orient using distinctive, difficult-to-erase features such as bridges, waterways, towers, and intersections.
- Leaders conduct navigational exercises in an actual urban area. Soldiers must move from point to point while controllers at each point monitor and verify Soldiers' progress.
- The GPS is a navigational aid that triangulates with satellites to calculate position. However, this system may be affected by nearby tall buildings, and its effectiveness decreases inside buildings or underground.

LIMITED VISIBILITY TRAINING

A-20. Urban operations involve day and night operations FTXs, STXs, CFXs, and CPXs are the most valuable exercises for night training.

- Field-expedient lighting, NVDs, illumination, and electrical lighting must be exploited when the tactical situation allows. Soldiers must learn discipline in their use of light, and also learn the lighting techniques that allow them to carry out their urban training program.

- Situational awareness is the Soldier's knowledge of his location, the locations of friendly forces, and the locations of the enemy on the battlefield. Situational awareness is the key to avoiding fratricide, especially during limited visibility operations. Units must plan to use all the equipment and devices available to mark friendly units and to avoid fratricide, without exposing themselves to enemy observation.
- Reverse-cycle training should be part of an urban training program. Limited visibility fighting skills are perishable, but they can be sustained and improved by effective training. The goal is for every Soldier to operate as effectively at night as during the day.

SAFETY AND AMMUNITION CONSIDERATIONS

A-21. When using ammunition and pyrotechnics, all associated safety requirements must be followed. This paragraph describes some of the safety equipment and training procedures to consider:

SPECIAL EFFECT SMALL-ARMS MARKING SYSTEM

A-22. SESAMS is 5.56-mm and 9-mm color-marking training ammunition. It is designed for close combat training. It is used with the close combat mission capability kit (CCMCK) adapter kit, which is installed on the M4/16 rifle, M9/M11 pistol, and M249 SAW. Different colors can be assigned to OPFORs for engagement identification and fratricide tracking. SESAMS has a 30-meter effective range and activates TES (MILES), so the two are used together. Anytime during a SESAMS exercise that a Soldier loses any of his protective gear, he must immediately cover his face and eyes with his hands. He must stay put until the engagement ends, or until he is told that it is safe to replace his safety equipment. Everyone, to include observer-controllers, must wear the protection shown in Table A-2.

Table A-2. SESAMS protection.

SKIN	Wear gloves, and cover all exposed skin. At close ranges, the SESAMS projectile can break skin.
HEAD	Take every precaution to protect your head.
EYES	Keep your eyes completely enclosed within eye protection. You can use standard Army sun, wind, and dust goggles.
EARS	Wear hearing protection to prevent hearing loss.
MOUTH	Keep your mouth covered.

DANGER

Failure to wear the required protection during SESAMS exercises can result in injury or death.

M84 STUN GRENADE

A-23. The hand-deployed M84 stun grenade serves as a reliable, effective, nonlethal means of neutralizing, diverting, and disorienting enemy personnel. It is used tactically as well as in training.

A-24. The M84 produces an intense (1 to 2.5-million candlepower) flash and, at 1.5 meters, a 170 dBs bang. Even though this is too low to produce permanent hearing damage, one-time use Soldiers are required to wear hearing protection anytime noise levels exceed 85 dBs.

A-25. The M84 stun grenade has an M201A1 fuze and a pyrotechnic output charge in a synthetic casing with a one- to two-second delay fuze, which cannot be cooked off. The grenade body has a nonfragmenting metal hexagonal casing, which, once deployed, resists rolling.

SMOKE GRENADES

A-26. Smoke grenades are used in urban combat to cover movement, deceive enemy forces as to the exact location of the attacking force, and for signaling. Smoke grenades should not be used in buildings and subterranean passages as they displace oxygen and protective masks do not filter the smoke or provide oxygen. Buildings in urban areas channel air currents, which influence emplacement of smoke grenades. One grenade should be used to test the airflow before employing additional grenades.

> **WARNING**
>
> **Avoid using smoke grenades in buildings and in subterranean passages.**

OPPOSING FORCES

A-27. OPFOR should be used as often as possible to enhance the realism of training, especially for FTXs, CFXs, and STXs. Thebe realistic, the OPFOR must have defined countertasks. These countertasks must support "Blue Force" training, and should prevent accidental contact between the Blue Force and the OPFOR. Leaders should develop a thorough control plan for OPFOR scenarios. This plan should describe uniform and vehicle markings, ammunition, safety instructions, controller guidance, and guidelines for handling enemy prisoners of war (EPW) and noncombatants. The ROE must clearly specify the limits of close combat between Soldiers.

- OPFOR heighten Soldiers' interest by giving them a real opponent. This fosters unit competition and realism, especially when the OPFOR is used in addition to TES and SESAM.
- OPFOR challenges leaders' ability to improvise. However, this free-play scenario should be monitored to ensure the unit meets the training objectives.
- OPFOR can use threat urban operations and doctrine to determine their course of action. If available, OPFOR uniforms, mock documents, overlays, and equipment add realism and intelligence to the scenario. Leaders conduct preliminary training for the OPFOR, which helps the OPFOR recreate the threat's urban tactics.
- Noncombatants should be integrated into training scenarios.

PRECISION MARKSMANSHIP TRAINING

A-28. Precision marksmanship training should be conducted before training at the UAC, CACTF, and the Shoot House, which accommodate training in target discrimination (FM 3-22.9). This is the best use of the unit's time and resources.

PHYSICAL READINESS TRAINING

A-29. Combat in urban areas places extraordinary physical demands on Soldiers. Therefore, physical readiness training, with emphasis on upper body development, is paramount to any urban operations training program.

VARIED SCENARIOS

A-30. Leaders must vary training scenarios to prevent redundant and predictable tactical behavior. Using predictable avenues of approach into the facility produces stereotypical tactical solutions.

BUILDING FACADE WALLS, WINDOWS, AND DOORS

A-31. The use of façade walls, windows and doors, in conjunction with SESAMS in the interior of some stations at the UAC, provides for more realistic training. This method of training reinforces the fact that walls, windows, and doors in most cases *do not* stop the penetration of service ammunition. Plastic wrap sheeting, tar paper, or some other type of material can be used to allow SESAMS penetration (Figure A-1).

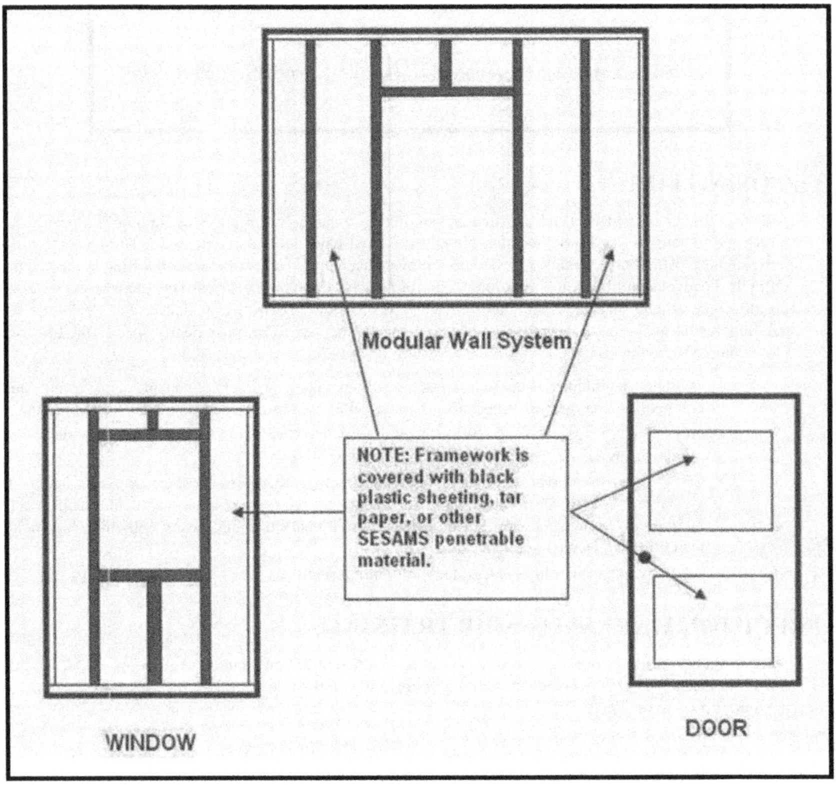

Figure A-1. Shoot-through walls and doors.

USE OF OLDER OR NONSTANDARD FACILITIES

A-32. Until installations receive funding for the UAC, SH, BF, and CACTF, they can use the older MOUT Assault Course (MAC) and Collective Training Facility (CTF) as described in the 2002 edition of TC 90-1, which this edition supersedes. Installations may also use nonstandard shoot houses and other nonstandard urban facilities. These facilities may be adapted as follows:

URBAN ASSAULT COURSE

A-33. Generally, the MAC can be used to train most of the UAC tasks. Of course, the MAC lacks targetry and instrumentation. Instead, units can record training with portable video cameras. During the AARs, they can play back the videos on VCRs hooked to TVs or on TV/VCR combos. OPFOR combatants will have SESAMS or MILES. People in civilian clothes can play the parts of the noncombatants.

SHOOT HOUSE

A-34. Many nonstandard shoot houses, such as tire houses, exist in different installations. Local range regulations and unit risk analyses will govern their use. Again, units can use portable video cameras for video capture and VCRs and television sets for video play back to facilitate an AAR. Both precision and nonprecision targets can be fabricated using E-type silhouettes on a stake, old BDUs/civilian clothes filled with straw, and so on. Simulate noncombatants with E-type silhouettes painted white or dummies in civilian clothes. Place balloons in the head and upper torso areas to simulate lethal shot placement. Where local policy permits, convert portions of collective training facilities into live-fire shoot houses using SRTA.

CAUTION

Units should ensure that they are in compliance with all local policies affecting live fire before attempting to use SRTA in an older collective training facility or nonstandard urban training site.

BREACH FACILITY

A-35. Existing demolition ranges may be used to conduct explosive breaching provided door, window, and wall units, or facsimiles, are available. Mechanical breaching can be trained in cantonment areas on the installation where building are designated for demolition, if such facilities are available. Regardless, whenever mechanical breacking is conducted, unit personnel must ensure that local safety policies are available, briefed, and complied with.

COMBINED ARMS COLLECTIVE TRAINING FACILITY

A-36. Generally the older CTF can be used to train the majority of tasks that would be trained in the CACTF. The CTF will not have any targetry and instrumentation. Again, units can use portable video cameras for video capture and VCRs and television sets for video play back to facilitate an AAR. Targets can be OPFOR with MILES. Role players in civilian clothes can be introduced to play the part of noncombatants. If armored vehicles are used, units should ensure that local range policies allow for their use. Older CTFs are normally either a 16 or 32 building design. Appropriate unit-level training, platoon, company, or battalion, should be planned based on the size of the facility.

Appendix B

Unit Training Matrix

Unit training plans are prepared IAW FM 7-1, which states that "Planning links the unit METL and the execution of battle-focused training.". The commander begins his planning with guidance from a higher echelon and an in-depth assessment of METL tasks (in this case, mission-essential urban tasks). Based on his own observations, feedback from his subordinates, and other sources the commander determines which tasks should be trained, which should be trained first, and how much time should be allotted for each. Tables B-1 through B-5 list training tasks by the WFF to be trained at each facility.

Table B-1. Army unit training matrix, brigade tasks.

Brigade Tasks	*SHOOT HOUSE Semiannual*	*BREACH HOUSE Semiannual*	*UAC Quarterly*	*CA/CTF Semiannual*	*CTC *As Scheduled*	*SIMULATION Annual*
Develop Intelligence						
Perform Intelligence Operations					X	X
Perform S-2 Operations					X	X
Deploy/Conduct Maneuver						
Attack					X	X
Perform Air Assault					X	X
Defend					X	X
Lodgment					X	X
Exercise Command and Control						
Command and Control the Brigade					X	X
Maintain Communications					X	X
Coordinate Air Defense Operation					X	X
Conduct Noncombatant Evacuation					X	X
Develop a Media Plan					X	
Conduct Mediation and Negotiation					X	
Employ Firepower						
Execute Fires					X	X
Employ Air Support					X	X
Employ Attack Helicopters					X	X
Synchronize Fire Support					X	X
Synchronize CAS					X	X
Conduct Targeting Process					X	X
Protect the Force						
Perform Mobility/Survivability Operations					X	X
Provide Engineer Support					X	X
Coordinate CBRN Operations					X	X
Conduct Operations Security					X	X
Perform Sustainment						
Evacuate Casualties					X	X
Treat Casualties					X	X
Sustain the Force					X	X
Conduct a Civil-Military Operation					X	X
Provide Humanitarian Support					X	X

Table B-2. Army unit training matrix, battalion tasks.

Battalion Tasks	*SHOOT HOUSE Semiannual*	*BREACH HOUSE Semiannual*	*UAC Quarterly*	*CA/CTF Semiannual*	*CTC As Scheduled*	*SIMULATION Annual*
Develop Intelligence						
Conduct S-2 Operations				X	X	X
Deploy/Conduct Maneuver						
Attack a Built-Up Area				X	X	X
Plan for Urban Operation				X	X	X
Defend an Urban Area				X	X	X
Conduct an Infiltration/Exfiltration				X	X	X
Perform Cordon and Search Operations in an Urban Area				X	X	X
Plan for Urban Operations				X	X	X
Conduct Tactical Movement				X	X	X
Conduct Presence Operation in a Stability Environment				X	X	X
Defend an Urban Area				X	X	X
Traverse a Built-up Area				X	X	X
Conduct a Mobile Defense				X	X	X
Employ Firepower						
Integrate Fires and Effects Support				X	X	X
Establish the Fire Support Element				X	X	X
Protect the Force						
Conduct Mobility or Survivability Operations				X	X	X
Secure Civilians During Operation				X	X	X
Conduct Passive Air Defense Measures				X	X	X
Conduct Active Air Defense Measures				X	X	X
Employ Operations Security Measures				X	X	X
Integrate Air Defense Support				X	X	X
Perform Sustainment						
Conduct S-1 Operations				X	X	X
Conduct S-4 Operations				X	X	X
Conduct a Civil-Military Operation				X	X	X
Exercise Command and Control						
Conduct S-6 Operations				X	X	X
Conduct Noncombatant Evacuation				X	X	X
Conduct S-3 Operations				X	X	X
Conduct Mediation and Negotiations				X	X	X
Plan Roadblocks or Checkpoints				X	X	X
Establish Liaison				X	X	X

Table B-3. Army unit training matrix, company tasks.

Company Tasks	SHOOT HOUSE Semiannual	BREACH HOUSE Semiannual	UAC Quarterly	CA/CTF Semiannual
Develop Intelligence				
Conduct an Area or Zone Reconnaissance				X
Reconnoiter a Built-Up Area				X
Deploy/Conduct Maneuver				
Conduct Cordon and Search				
in a Built-Up Area				X
Attack a Built-Up Area				X
Conduct Presence Operations				X
Conduct Tactical Movement				
in a Built-Up Area				X
Defend in Urban Environment				X
Conduct an Infiltration (Co/Plt)				X
Conduct Operations with Armored or				
Mechanized Infantry Vehicles				
in an Urban Environment				X
Conduct Civil Disturbance Operations				X
Establish Checkpoints				X
Employ Firepower				
Integrate Direct Fires				X
Integrate Indirect Fires				X
Protect the Force				
Maintain Operations Security				X
Conduct Passive Air Defense Measures				X
Conduct Passive Air Defense Measures				X
Perform Sustainment				
Treat Casualties				X
Evacuate Casualties				X
Secure Civilians during Operations				X
Conduct Resupply Operations				
Process Captured Documents and				X
Equipment				X
Handle Enemy Prisoners of War				
Exercise Command and Control				
Prepare for Combat				X
Establish Communications				X
Conduct Consolidation and Reorganization				X
Conduct Negotiation				X
Conduct Troop-Leading Procedures				X

Table B-4. Army unit training matrix, platoon/squad tasks.

Platoon/Squad Tasks	*SHOOT HOUSE Semiannual*	*BREACH HOUSE Semiannual*	*UAC Quarterly*	*CA/CTF Semiannual*
Develop Intelligence				
Report Tactical Information	X		X	X
Conduct an Area or Zone Reconnaissance	X		X	X
Reconnoiter a Built-Up Area	X		X	X
Deploy/Conduct Maneuver				
Conduct Tactical Movement in a Built-Up Area	X		X	X
Assault a Building	X		X	X
Conduct a Strongpoint Defense of a Building				X
Conduct an Infiltration or Exfiltration				X
Conduct Operations with Armored or Mechanized Vehicles in an Urban Environment	X			X
Conduct a Presence Patrol	X			X
React to a Civil Disturbance				X
Establish a Checkpoint				X
Search a Building	X		X	X
Employ Firepower				
Employ Fire Support				X
Protect the Force				
Maintain Operations Security				X
Conduct Passive Air Defense Measures				X
Conduct Active Air Defense Measures				X
Perform Sustainment				
Secure Civilians During Operations	X		X	X
Perform Resupply Operations				X
Treat Casualties	X		X	X
Evacuate Casualties	X		X	X
Handle Enemy Prisoners of War				
Exercise Command and Control				
Conduct Troop-Leading Procedures	X		X	X
Establish Radio Communication	X		X	X
Conduct Consolidation and Reorganization	X		X	X
Prepare for Combat	X		X	X

Table B-5. Army unit training matrix, individual tasks/techniques.

Individual Tasks/Techniques	*SHOOT HOUSE* *Semiannual*	*BREACH HOUSE* *Semiannual*	*UAC* *Quarterly*	*CA/CTF* *Semiannual*
Skill Level 1				
Perform Movement Techniques in an Urban Environment, 071-326-0541	X	X	X	X
Prepare Positions for Individual and Crew-Served Weapons in an Urban Environment, 071-326-0550	X	X	X	X
Select Hasty Fighting Positions in an Urban Environment, 071-326-0557	X	X	X	X
Employ Hand Grenades, 071-325-4407	X	X	X	X
Engage Targets with an M16-Series or M16A2 Rifle, 071-311-2007	X	X	X	X
Execute a Mechanical Breach, 071-440-0027			X	X
Skill Level 2				
Conduct an Explosive Breach	X	X	X	X
Conduct a Breach, 071-440-0029	X	X	X	X
Prime Explosive Nonelectrically, 051-193-1003	X	X	X	X
Construct a Nonelectric Initiating Assembly with Modernized Demolition	X	X	X	X

Appendix C

Stability and Support Operations Unit Training Matrix

When assigned a stability or support mission, a well-trained unit must be able to rapidly shift its focus from war fighting to stability and support and also from stability and support to war fighting. During a stability or support operation, the unit performs numerous activities. Essentially, the unit accomplishes these activities through execution of tactical missions and tasks. While stability and support operations can occur anywhere, it is most probable that they will occur in an urban environment. Tables C-1 through C-4 list training tasks by the WFF to be trained at each facility.

Table C-1. Stability and support operations unit training matrix, brigade tasks.

Brigade Tasks	SHOOT HOUSE	BREACH HOUSE	UAC	CA/CTF	CTC	SIMULATION
Develop Intelligence						
Collect and Disseminate Information					X	X
Deploy/Conduct Maneuver						
Demonstrate a Show of Force					X	
Employ Psychological Operation					X	X
Employ Firepower						
Provide Fire Support					X	X
Protect the Force						
Protect the Force					X	
Perform Sustainment						
Restore Law and Order					X	
Exercise Command and Control						
Establish an Evacuation Control Center/Conduct Evacuation Control Center Operations					X	
Negotiate					X	
Perform Noncombatant Evacuation Operations					X	
Plan a Media Visit					X	
Develop Intelligence						
Collect and Disseminate Information				X	X	X

Table C-2. Stability and support operations unit training matrix, battalion tasks.

Battalion Tasks	*SHOOT HOUSE*	*BREACH HOUSE*	*UAC*	*CA/CTF*	*CTC*	*SIMULATION*
Deploy/Conduct Maneuver						
Interdict Smuggling Operations				X	X	
Disarm Belligerents				X	X	
Handle Captured Insurgents or Belligerents				X	X	
Employ Psychological Operation				X	X	
React to Civil Disturbance				X	X	
Employ Firepower						
Provide Fire Support				X	X	X
Protect the Force						
Protect the Force				X	X	
Restore Law and Order						
Identify and Process Detainees				X	X	
Deliver Supplies or Humanitarian Aid				X	X	
Control Civilian Movement				X	X	
Prepare Traffic Control Plan				X	X	
Exercise Command and Control						
Negotiate				X	X	
Perform Noncombatant Evacuation Operations				X	X	
Conduct Marshalling Force Operations				X	X	
Negotiate a Belligerent Force Checkpoint				X	X	
Plan a Media Visit				X	X	
Perform a Cordon and Search				X	X	
Coordinate and Monitor Convoy Security				X	X	

Table C-3. Stability and support operations unit training matrix, company tasks.

Company Tasks	SHOOT HOUSE	BREACH HOUSE	UAC	CA/CTF	CTC	SIMULATION
Develop Intelligence						
Collect and Disseminate Information				X	X	
Deploy/Conduct Maneuver						
Interdict Smuggling Operations				X	X	
Disarm Belligerents				X	X	
Handle Captured Insurgents or Belligerents				X	X	
Defend Convoy				X	X	
Apprehend/Detain Noncombatants				X	X	
Perform Cordon and Search				X	X	
Employ Firepower						
Provide Fire Support				X	X	
Protect the Force						
Establish a Checkpoint				X	X	
Operate a Checkpoint				X	X	
Protect the Force				X	X	
Perform Sustainment						
Process Captured Documents and Equipment				X	X	
Exercise Command and Control						
Negotiate a Belligerent Force Checkpoint				X	X	
Conduct Marshaling Force Operations				X	X	
Negotiate a Belligerent Force Checkpoint				X	X	
Plan a Media Visit				X	X	
Coordinate and Monitor Convoy Security				X	X	
Link Up with a Convoy				X	X	

Table C-4. Stability and support operations unit training matrix, platoon/squad tasks.

Platoon/Squad Tasks	*SHOOT HOUSE*	*BREACH HOUSE*	*UAC*	*CA/CTF*	*CTC*	*SIMULATION*
Deploy/Conduct Maneuver						
Disarm Belligerents				X	X	
Handle Captured Insurgents or Belligerents				X	X	
Apprehend/Detain Noncombatants				X	X	
Search a Building	X			X	X	
React to a Sniper			X	X	X	
Conduct Platoon Riot Control Formation				X	X	
Protect the Force						
Establish a Checkpoint				X	X	
Operate a Checkpoint				X	X	
Protect the Force				X	X	
Exercise Command and Control						
Negotiate a Belligerent Force Checkpoint				X	X	
Exercise Command and Control						
React to a Media Interview				X	X	

Appendix D
Leader Development Training

To be successful in combat, the Army must continually train to develop and maintain combat-ready Soldiers, leaders, and units that can perform assigned tasks to specific standards.

SIMULATIONS TRAINING

D-1. Commanders train their staffs to plan, coordinate, and synchronize their units' operations from initial mobilization and deployment. These operations include the full spectrum of Army operations; offense, defense, stability, and support. Current simulations lack adequate replication of urban scenarios, terrain, and weapons effects. The introduction of WARSIM 2000 and OneSAF should help to solve these simulations modeling deficiencies. For lower echelon units, battalion and below, constructive simulations complement collective battle staff training that occurs during virtual simulation similar to the close combat tactical trainer and live events. For echelons above battalion, it becomes difficult to use simulators or live maneuver to train. Studies to date indicate the crossover point for simulation versus simulator is at the battalion level. Some training events from battalion and below can be trained effectively using various types of simulators. Above battalion level, effective training almost has to be conducted with a simulation (Figure D-1).

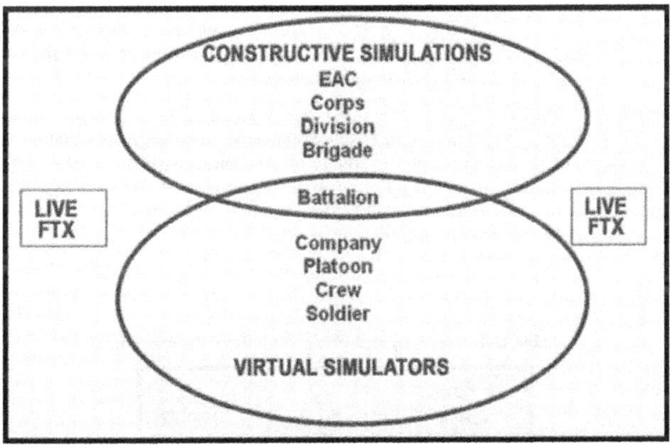

Figure D-1. Echelons and uses.

D-2. Simulations that support leader and battle staff training focus on collective tasks rather than individual leader tasks. Simulations, used in this context, rely on the human dynamics in the command setting and use real-time actions in a tactical environment with a unit's go-to-war systems. This process forces commander and staff interaction that provides feedback and helps refine the integration process. Battle staff training encourages team building, task coordination, and standardization of staff actions.

- Constructive simulations can provide commanders large areas for conducting training, access to sensitive areas through computer and mapping technology, and futuristically provide a much better level of fidelity in terms of urban operations terrain and scenario replication.
- Simulations can portray large, capable, and doctrinally correct OPFORs. All of the Soldiers portrayed in the units in a simulation do not have to be present in order for the training to be realistic. An attractive feature of simulations has always been the reduced staffing necessary to conduct this type of training exercise.
- Simulations can stress CPs and provide the realistic conditions under which commanders must make decisions. Brigade and below simulation (BBS) and corp battle simulation (CBS) are well known for the stressful environments created during an exercise. The stress is realistic and can assist in conditioning the training audience before CTC rotations.
- Simulations allow different units to train under the same conditions and to the same standards. This technique is a unique feature of simulations as the exact weather, opposing force organization, location of units, and equipment can be used to train many different units on the same terrain with the same conditions.
- The size and number of personnel in a battle staff vary depending on the echelon and other demands unique to that commander. However, the WFF are represented by the following staff officers: intelligence, operations, fire support, communications, logistics, air defense, and engineer.
- Characteristics and benefits of simulations used to support battle staff training follow:
  - -- Build and sustain staff teamwork.
  - -- Build and sustain effective and efficient organizations.
  - -- Provide training and validation of command and staff procedures (SOPs, TACSOPs, FSOPs).
  - -- Provide proficiency to the individual staff officer, while building proficiency in a fully integrated staff.
  - -- Allow practice of battle drills and tactical missions (movement to contact, hasty defense).
  - -- Allow commanders to determine if their staff and subordinate commanders understand the commander's intent and concept of operations.

D-3. The major development efforts for FAMSIM is WARSIM 2000. WARSIM 2000, in conjunction with OneSAF, exploits new technologies to enable commanders and staffs at battalion through EAC to train in a realistic simulation environment (Table D-1). The increased realism of WARSIM 2000 over existing models allows units to use Army Battle Command Systems (ABCS) to synchronize across the complete range of operations. WARSIM 2000's design allows warfighting CPs to use organic ABCS equipment to interact with the simulation from training sites in the field, as well as from simulation centers. WARSIM 2000 will also be capable of showing a joint and combined environment across the operational continuum.

Table D-1. Urban operations simulations crosswalk.

CAMTF Training Tasks	*WARSIM*	*OneSAF*
Corps	X	
Division	X	
Brigade		X
Battalion		X
Company		X
Platoon		X

D-4. Future simulation capabilities simultaneously integrate live, virtual, and constructive environments. Constructive simulation is a viable solution for battalion and above training and the federation of WARSIM and OneSAF models should alleviate the following training simulation problems:

- Eliminate the need for swivel chair interface.
- Allows focus on subordinate units fighting the fight.
- Digital and analog feeds to C4I systems including FBCB2.
- May provide for distance learning opportunities.
- Facilitates simultaneous multi-echelon urban operations training

D-5. At company and below, the best training solution is live and virtual. This is accomplished by linking the Close Combat Tactical Trainer (CCTT) to OneSAF and VLET to Engagement Skills Trainer (EST) technology. Entity-based models like OneSAF can allow the leader to fight from a TOC or vehicle, if linked to virtual or live environments. It also has the potential for distance learning.

PHASES OF TRAINING

D-6. The Army Urban Operations Training Strategy consists of initial, intermediate, and culminating stages. These phases of training, shown in Figure D-2, describe the sequence in which urban operations training tasks should be conducted.

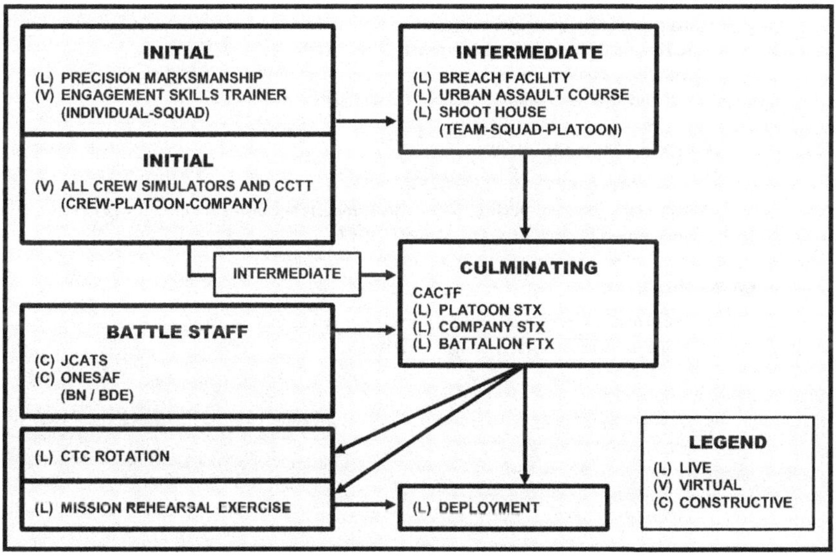

Figure D-2. Phases in the Live, Virtual, and Constructive Training Strategy.

Initial Phase

D-7. The initial phase of training covers those individual and collective tasks that Soldiers and leaders should be proficient at before arriving at any of the urban operations training sites. They consist primarily of basic marksmanship skills and Soldier tasks needed to perform any tactical operation.

Intermediate Phase

D-8. The intermediate phase consists of training conducted at the breach facility, shoot house, and the urban assault course. Urban operations training tasks should be conducted here before advancing to the Combined Arms Collective Training Facility with the exception of Aviation, Armor, and other branches that have no practical use for the Shoot House, Breach Facility, or Urban Assault Course.

Culminating Phase

D-9. The combined arms collective training facility is the culminating phase of urban operations training and focuses on multiechelon collective skills for all branches of the combined arms team. Urban operations leader training currently exists in our leader professional development courses. Units have developed internal urban warfare training programs and TRADOC schools have assembled technical mobile training teams (MTTs) for marksmanship, combatives, and breaching techniques.

Education Phase

- Urban warfare reading/video list.
- Urban operations equipment.
- Review of doctrinal publications (FM 3-21.10, FM 3-06.11, for Infantry) and corresponding publications for other branches.
- Small-unit urban movement techniques and considerations.

Planning and Execution Phase

- Prepare company OPORD or WFF-related products.
- Brief platoon/company OPORD/WFF-related products to commander or staff representative.
- Participate as briefer in urban environment TEWT.

Sustainment Phase

- Participate in company/platoon TEWT or other urban operations event.
- Execute/facilitate a company or platoon TEWT.

EXAMPLE LEADER'S COURSE POI

D-10. An example urban leaders' course POI, which is intended to provide unit trainers with a tailorable urban operations training program, is shown in Table D-2. It is a *train-the-trainer* course for team leaders through platoon leaders to train tactics, techniques, and procedures of advanced urban operations skills. The length and focus of the POI can be tailored to support the commander's training objectives and unit METL. The target audience is newly arrived, or soon to be, small unit leaders. The recommended course length is about two weeks or 120 to 140 hours of instruction.

REALISTIC TRAINING OFF FEDERAL FACILITIES

D-11. Realistic training for urban operations is critical to force readiness. Training opportunities that are both realistic and challenging may be unavailable at federal installations, or at facilities specifically designed for urban operations training. Training in off-post civilian urban settings may become necessary in order to support readiness and training requirements.

D-12. The Department of Defense has developed the following SOP for planning and approving such events. The SOP assists in balancing the needs of the military training and of the civilian community/ies to be used for realistic urban training events.

Table D-2. POI for urban leader's train-the-trainer course.

Course Overview	• Introduction	
Precision Marksmanship	• Firing techniques. • Target discrimination.	
Breaching Techniques	• Explosive. • Ballistic. • Mechanical.	

Table D-2. POI for urban leader's train-the-trainer course (continued).

Offensive Operations	• Planning considerations	-- METT-TC. -- Snipers. -- CASEVAC. -- Communications. -- ROE.
	• Platoon attack a built-up area (task organization)..	-- Assault (isolate, enter, clear). -- Rooms, hallway, stairs, superstructure, subsurface. -- Markings. -- Movement techniques. -- Streets. -- Alley ways. -- Obstacles. -- Other danger areas. -- Open areas. -- Roof tops.
	• Combined arms	-- Armor/mechanized platforms. -- Aviation assets. -- Artillery. -- Engineer.
	• Limited visibility	
Defensive Operations	• Planning considerations	-- METT-TC. -- Snipers. -- CASEVAC. -- Communications. -- ROE.
	• Characteristics of urban areas	
	• Hasty defense	-- Occupation and preparation of positions. -- Hasty firing positions. -- Improving the defense.
	• Deliberate defense, including priorities of work and other considerations	
	• Defensive plan at platoon level.	-- Defense of a strongpoint. -- Defense against armor.
	• Combined arms	-- Armor/mechanized. -- Aviation assets. -- Artillery. -- Engineer.
	• Limited visibility • Fundamental combat skills	

Table D-2. POI for urban leader's train-the-trainer course (continued).

Stability and Support Operations	• Principles of stability and support.	
	• Stability and support activities	-- NEO. -- Humanitarian assistance. -- Disaster relief. -- Show of force. -- Attack and raids.
	• Rules of engagement application • Convoy operations • Checkpoints • Fire support • Sniper employment/counter sniper measures • Civil disturbance	
Urban Training Facilities	• Breach facility • Urban assault course • Shoot house • CACTF	
New Technologies	• Rifle launch entry munition	
	• Unmanned vehicles	-- UAV. -- UGV. -- Ladders. -- Night vision devices.

STANDING OPERATING PROCEDURES FOR PLANNING AND APPROVING REALISTIC URBAN TRAINING

1. **REFERENCES.**

   10 USC § 167, *Unified Combatant Command for Special Operations Forces.*

   10 USC §§ 3013, 5013, 8013.

   Department of Defense Directive 5111.10.

   Unified Command Plan 99.

2. **PURPOSE**. This SOP establishes uniform planning and approval procedures for realistic urban training (RUT) events.

3. **APPLICABILITY**. This SOP applies to all RUT events in civilian urban settings within the United States, its territories, and its possessions; and to Active or Reserve Forces, including National Guard Forces in Federal service. This policy does not apply to National Guard Forces under in-state status (Title 32), or to aviation operations that do not require an exemption from standard Federal Aviation Regulations. The training event should not be located on federal state, local, or private installations or facilities.

4. **DEFINITIONS**. For the purposes of this SOP, the following definitions apply:

a. **Urban Operations. This term** includes (see FM 3-06) all military actions planned and conducted on a topographical complex, and on adjacent natural terrain, where man-made construction is the dominant feature. It includes combat in towns and cities, including house-to-house and street-by-street fighting.

b. **Realistic Urban Training.** RUT is high-intensity, close-quarter battle training using live or nonlethal fires; demolitions or explosives such as breaching charges; and air and naval supporting platforms at the objective, in ***civilian urban settings***.

c. **Civilian Urban Setting.** A civilian urban setting is any area(s) located in or near a civilian community. Civilian urban settings will not be developed or established for training events, especially where training activities would affect civilian residences or commercial areas due to noise, traffic, and so on.

5. **PROCEDURES**. The following procedures apply to the planning and approval of RUT in civilian urban settings:

a. The combatant commander or the service ensures compliance with this SOP. Approval authority for RUT events may be delegated to a general or flag officer within the chain of command, or to the combatant common service conducting the training.

b. The approval authority reviews proposed training events to determine whether they are necessary and appropriate for unit readiness. Training must support unit METs and JMETs.

c. On approval of the proposed training, and at least 14 days before training starts, the combatant common service notifies the **SECR/PF** through the CJCS (by way of message) of the approved training. At a minimum, this notice includes the following:

- Name and location of the unit conducting training.
- Scope of training and activities planned such as live fire, breaches, or use of aircraft.
- Start and end dates.
- Training location.
- A statement that all necessary coordination has been accomplished IAW this SOP. The statement should include a list of the contacted official's names and their positions.

d. **Coordination with Civilian Officials**.

- While planning the training event, the commander coordinates with and obtains written approval to conduct the event from the appropriate federal, state, and local civilian authorities. Commanders will consult their legal and public affairs officers to identify the civilian officials, for example the mayor, borough chief, county commissioners or supervisors, with whom they must coordinate the training event. To do this, they consult local and Federal law enforcement agencies. At a minimum, they must consult a senior official with responsibility for each affected civilian urban setting.
- During coordination, the commander covers the details of the proposed training activities, such as whether the event involves live fire, rotary- or fixed-wing aircraft, or close quarters battle training; and whether the exercises will be conducted during the day, night, or both: and so on. It will also list actions to ensure public and participant safety, proposed public or resident notification and outreach procedures, and any other issues that might cause public interest or concern.
- As much as possible, the commander briefs civilian officials in person on the exercise concept.

- The US congressional representative or staff for the training district, and the US senators for the state, or their staffs, are advised of the proposed training event and given whatever information they request. Neither written approval nor concurrence from the representative or senators is required. However, if these officials raise objections or issues, and if those issues cannot be resolved, they are immediately elevated to the Joint Staff, J3, and the service or combatant commander point of contact.
- For each training site, license agreements with civilian or government property owners or managers are executed. Agreements address liability and conditions of acceptable usage of property. Property owners are informed of the nature of the training including written notification as to whether the activities will include use of breaching charges and, if so, how much; or live ammunition and, if so, which weapons and what ammunition. This notice will also state *if* rotary or fixed-wing aircraft will be used *and* whether the exercises will be conducted during the day, night, or both. Before the exercise can begin, all license agreements must be reviewed for legality. Copies of the license agreements will be provided to the senior civilian officials responsible for training area.
- A record of coordination activities is maintained for each training event. The record includes the names and positions of the officials *who* coordinate and approve the event. It also provides officials with a summary of the events such as date, time, and location of the training; types of activities; and planned public notification and outreach. A follow-on letter is distributed to the senior civilian officials responsible for the affected civilian urban settings. The letter confirms the training event *and* lists the names of the civilian officials who plan the training event.
- The unit may schedule recurring training events at a particular area or location by submitting a single MOA to the authorizing civilian officials. The terms of the MOA must comply with the requirements of this SOP. The service or the combatant commander reviews the MOA, which authorizes the recurring training, to ensure it complies with the procedures. Once the commander approves the MOA, the unit may conduct individual training events under it without further Joint Staff or OSD review. However, services and combatant commanders notify SO/LIC *and* the Joint Staff 14 days *before* the training in this MOA begins. Thereafter, services and combatant commanders must review these MOAs periodically to ensure *that* they continue to fulfil the intent.
- *If* nondisclosure agreements are required, command legal representatives must review them. Nondisclosure statements should clearly explain *that* the agreement does not preclude persons from informing others of general, nonsensitive information about the training exercise.
- Depending on OPSEC and safety considerations, commanders must adhere to the recommendations of civilian officials regarding community notification, outreach, and press procedures. Community notification and outreach procedures may include door-to-door notification by the local law enforcement agency. Commanders will advise civilian officials when they cannot follow their recommendations.

e. Other instructions include--

- Legal and PA representatives attend final planning meetings, and they remain on-site throughout all training activities.
- The on-site PAO handles all press queries after coordinating with the on-scene commander.
- All requested training must follow applicable statutes, law, DOD directives, other guidance, and implementing service directives.

Glossary

Section I. ACRONYMS AND ABBREVIATIONS

1SG first sergeant

A

AA assembly area

AAR after-action review/report

ABCS Army Battle Command System

ACU Army combat uniform

ADA air defense artillery

ANCOC advanced noncommissioned officer course

AO area of operation

AR Army Reserve

AT4 a shoulder-fired munition, formerly called a "light antiarmor weapon"

atk attack

ATWESS antitank weapon effect signature simulator

AVCATT armored vehicle combined arms tactical trainer

B

BBS brigade and below simulation

BCT brigade combat team

bde brigade

BDU (obsolete; *see* ACU)

BF Breach Facility

bn battalion

BNCOC basic noncommissioned officer course

BOLC Basic Officer Leadership Course

BOS (obsolete; *see* WFF)

BP battle position

BSA brigade support area

BSX battle simulation exercise

C

C2 command and control

CACTF Combined Arms Collective Training Facility

CASEVAC casualty evacuation

CBRN chemical, biological, radiological, nuclear

CBS corps battle simulation

CCMCK close combat mission capability kit

CCOC Captain's Career Officer Course

CCTT close combat tactical trainer

CCTV closed-circuit television

CFX command field exercise

CMTC Citizens Military Training Corps

co company

commo communications

CP command post

CPX command post exercise

CTC Combat Training Center

CTF collective training facility

D

DES demolition effects simulator

det cord detonation cord

div division

DLIC detachment left in contact

DS Digital Training Management System

DTMS direct support

E

EA engagement area

EAC echelons above corps

ENCATT engineer combined arms tactical trainer

engr engineer

EPW enemy prisoners of war

EST Engagement Skills Trainer

F

FAMSIM family of simulations

FO forward observer

FRAGO fragmentary order

FSCATT fire support combined arms tactical trainer

FSOP field standing operating procedures

FTX field training exercise

G

GREM	grenade rifle entry munition
GSR	ground surveillance radar

H

HC	herachloroethane zinc
HEAT	high-explosive antitank

I

IAW	in accordance with
Inf	Infantry

J

JANUS	Joint Army-Navy Uniform Simulation
JCATS	Joint Conflict and Tactical Simulation
JMET	joint mission-essential task
JRTC	Joint Readiness Training Center
JTF	joint task force
JTX	joint training exercise

L

LAW	light antiarmor weapon
LFX	live-fire exercise
LMG	light machine gun

M

MAC	MOUT assault course
MACOM	major command
MAPEX	map exercise
M/C/S	mobility/countermobility/ survivability
MET	mission-essential task
METL	mission-essential task list
METT-TC	mission, enemy, terrain, troops, time available, and civilians
MI	military intelligence
MILES	Multiple Integrated Laser-Engagement System
MOPP	mission-oriented protective posture
MPAT	multipurpose antitank (ammunition)
MPAT-OR	multipurpose antitank-obstacle reducing
MRE	meals ready to eat
MTT	mobile training team

N

NBC	(obsolete; *see* CBRN)
NEO	noncombatant evacuation operations
NLT	not later than
NTC	National Training Center

O

OneSAF	one semiautomated force (software used to create modeling and simulation programs with military applications)
op	operation
OPFOR	opposing force
OPORD	operation order
OPSEC	operations security

P

plt	platoon
POI	program of instruction
PPP	power projection platform
PSP	power support platform

R

ROE	rules of engagement
ROI	rules of interaction
RUT	realistic urban training

S

S-1	personnel staff officer (adjutant)
S-2	intelligence staff officer
S-3	operations staff officer
S-4	logistics staff officer
S-5	civil affairs officer
S-6	command, control, communications and computer operations (C4 operations) officer
SAAF	small-arms alignment fixture
SBCT	Stryker brigade combat team
SDZ	surface danger zone
SEP	system enhancement program: a version of the M1A2 machine gun with enhanced digital command and control and second-generation forward-looking infrared

SESAMS	Special Effects Small Arms Marking System
SH	Shoot House
SOI	signal operation instructions
SOP	standing operating procedure
sqd	squad
SRTA	short-range training ammunition
STAFFEX	staff exercise
STX	situational training exercise

T

T&EO	training and evaluation outline
TACSOP	tactical standing operating procedure
TES	Tactical Engagement System
TEWT	tactical exercise without troops
TF	task force
TOW	tube-launched, optically tracked, wire-guided (missile)
TTP	tactics, techniques, and procedures

U

UAC	urban assault course
UO	urban operations

V

VCR	videocassette recorder
VLET	virtual leader effects trainer

W

WARSIM	warfighting simulation
WFF	warfighting function
WFX	warfighting exercise

X

XO	executive officer

Section II. TERMS

bustle rack storage rack mounted on turrets of some armored vehicles

entity-based model a representation of an actual or conceptual system that includes mathematical, logistical, or computer simulation

warfighting functions group of tasks and systems (people, organization, information, and processes) united by a common purpose that commanders use to accomplish missions and training objectives. The warfighting functions are intelligence, movement and maneuver, fire support, protection, sustainment, and command and control. These warfighting functions replace the battlefield operating systems (BOSs).

References

SOURCES USED/DOCUMENTS NEEDED

These are the sources quoted or paraphrased in this publication. All of them are needed by and must be available to the intended user.

DOD DIRECTIVE

DOD Directive 2311.01E *DOD Law of War Program.* 9 May 2006.

FIELD MANUALS

FM 3-06	*Urban Operations.* 26 October 2006.
FM 3-06.11	*Combined Arms Operations in Urban Terrain.* 28 February 2002.
FM 3-20.12	*Tank Gunnery (Abrams).* 15 August 2005.
FM 3-21.8	*The Infantry Platoon and Squad.* 28 March 2007.
FM 3-21.9	*SBCT Infantry Rifle Platoon and Squad.* 2 December 2002.
FM 3-21.10	*The Infantry Rifle Company.* 27 July 2006.
FM 3-21.11	*The SBCT Infantry Rifle Company.* 23 January 2003.
FM 3-21.20	*The Infantry Battalion.* 13 December 2006.
FM 3-21.21	*The Stryker Brigade Combat Team Infantry Battalion.* 8 April 2003.
FM 3-21.71	*Mechanized Infantry Platoon and Squad (Bradley).* 20 August 2002.
FM 3-22.9	*Rifle Marksmanship M16A1, M16A2/3, M16A4, and M4 Carbine.* 24 April 2003.
FM 7-0	*Training the Force.* 22 October 2002.
FM 7-1	*Battle Focused Training.* 15 September 2003.
FM 25-4	*How to Conduct Training Exercises.* 10 September 1984.

SOLDIER TRAINING PUBLICATIONS

STP 21-1-SMCT	*Soldier's Manual of Common Tasks Skill Level 1.* 14 December 2007.
STP 21-24-SMCT	*Soldier's Manual of Common Tasks (SMCT) Warrior Leader Skill Levels 2-4.* 2 October 2006.

TRAINING CIRCULAR

TC 7-98-1	*Stability and Support Operations Training Support Package.* 5 June 1997.

Index